NCA Report Series, Volume 5c

Climate Change Impacts and Responses: Societal Indicators for the National Climate Assessment

NCA Report Series

The National Climate Assessment (NCA) Report Series summarizes regional, sectoral, and process-related workshops and discussions being held as part of the Third NCA process.

The workshop on including and developing societal indicators as a part of the 2013 NCA was held in Washington, DC on April 28-29, 2011. Volume 5c of the NCA Report Series summarizes the discussions and outcomes of this workshop. A list of completed and planned reports in the NCA Report Series can be found online at http://assessment.globalchange.gov.

CONTENTS

Executive Summary

The Climate Change Impacts and Responses: Societal Indicators for the National Climate Assessment workshop, sponsored by the National Aeronautics and Space Administration (NASA) and the National Oceanic and Atmospheric Administration (NOAA) for the National Climate Assessment (NCA), was held on April 28-29, 2011 at The Madison Hotel in Washington, DC. A group of 56 experts (see list in Appendix B) were convened to share their experiences. Participants brought a wide range of disciplinary expertise in the social and natural sciences, sector experience, and knowledge about developing and implementing indicators for a range of purposes. Participants included representatives from federal and state government, nongovernmental organizations (NGOs), tribes, universities, and communities.

The purpose of the workshop was to assist the NCA in developing a strategic framework for climate-related physical, ecological, and socioeconomic indicators that can be easily communicated with the U.S. population and that will support monitoring, assessment, prediction, evaluation, and decision making. The NCA indicators are envisioned as a relatively small number of policy-relevant integrated indicators designed to provide a consistent, objective, and transparent overview of major variations in climate impacts, vulnerabilities, adaptation, and mitigation activities across sectors, regions, and timeframes.

The workshop participants were asked to provide input on a number of topics, including (1) categories of societal indicators for the NCA; (2) alternative approaches to constructing indicators and the better approaches for NCA to consider; (3) specific requirements and criteria for implementing the indicators; and (4) sources of data for and creators of such indicators. Socioeconomic indicators could include demographic, cultural, behavioral, economic, public health, and policy components relevant to impacts, vulnerabilities, and adaptation to climate change as well as both proactive and reactive responses to climate change.

Participants provided inputs through in-depth discussion in breakout sessions, plenary sessions on break-out results, and several panels that provided key insights about indicators, lessons learned through experience with developing and

implementing indicators, and thoughts on how the NCA could proceed to develop indicators (see Agenda in Appendix A).

Breakout groups were charged with addressing questions related to four main themes over the two-day workshop: (1) NCA indicator framework goal, audience, and scope; (2) benefits or drawbacks and lessons learned of different indicator approaches; (3) "must have" topical societal indicator categories; (4) categories, requirements, data, and priorities for developing climate impacts indicators, climate adaptation indicators, climate vulnerability and resiliency indicators, and climate disaster preparedness indicators; and (5) general recommendations on developing societal indicators for the NCA.

During the workshop discussions, a number of points emerged as key messages worth considering as the NCA moves forward in developing an indicator framework:

- Indicators developed or selected for the NCA should motivate the audience to notice and pay attention (be relevant to topics they care about), believe the information (because it is credible), and do something about it (because it is actionable).
- The NCA should start with the questions to be answered and then choose the indicators to best address the question.
- The NCA should draw lessons from and, where appropriate, build upon the many other indicators and indicator approaches that have been developed to address similar issues, as reviewed in the workshop. The indicator approach (e.g., composite, basket, and accounting) does not need to be the same for all of the indicator categories.
- The NCA should start with what is doable (i.e., "low hanging fruit"), especially in the short-term, and leverage existing efforts when possible.
- Indicators developed or selected for the NCA should be scientifically defensible, meet NCA peer-review standards, and be transparently presented in message, approach, and data sources.
- The NCA should engage stakeholders early and often in a two-way conversation, remembering that not all stakeholders are the same.
- The NCA indicator framework should be flexible, customizable, and serve multiple audiences in a way that builds common

understanding among different groups.
- The process for selecting and developing indicators could include "citizen science" and experiential knowledge approaches.
- The indicators developed or selected for the NCA should be representative, not comprehensive (especially in the short-term).
- The indicators need to have appropriate coverage and be consistently gathered.
- The indicators developed or selected for the NCA should reflect both negative and positive aspects of climate (i.e., impacts and opportunities, vulnerabilities and resiliencies).
- The indicators selected should have enough frequency and consistency to be measured over time.
- The indicators developed or selected for the NCA should be evaluated and adaptively managed to allow for changes over time.

Part I:
Workshop Report – Summary of Presentations and Discussions

Written by: Melissa A. Kenney, Robert S. Chen, Julie Maldonado,
and Dale Quattrochi

1 OVERVIEW OF THE WORKSHOP

This part of the workshop report summarizes the presentations and discussions that occurred at the workshop "Climate Change Impacts and Responses: Societal Indicators for the National Climate Assessment" (referred to as the Societal Indicators workshop) on April 28-29, 2011, sponsored by the National Aeronautics and Space Administration (NASA) and the National Oceanic and Atmospheric Administration (NOAA) for the National Climate Assessment (NCA). The purpose of the workshop was to assist the National Climate Assessment (NCA) in developing a strategic framework for climate-related physical, ecological, and societal indicators that can be easily communicated with the U.S. population that will support monitoring, assessment, prediction, evaluation, and decision making. The NCA indicators are envisioned as a relatively small number of policy-relevant integrated indicators designed to provide a consistent, objective, understandable, and transparent overview of major variations in climate impacts, vulnerabilities, adaptation, and mitigation activities across sectors, regions, and timeframes. Over 50 people participated in the workshop, including social science researchers with expertise in

- best practices for developing indicators,
- indigenous cultures and Tribes,
- poverty and social vulnerability,
- adaptive capacity,
- multi-stakeholder decision making,
- environmental governance and institutions,
- environmental justice and equity,
- complex emergencies and disasters,
- food security and agricultural development,
- land and water resource management,
- energy security,
- economic development and growth, and
- remote sensing data as applied to human health and societal impacts.

The program was developed with input from the workshop steering committee, which included representatives of the social science community. The workshop and steering committee were chaired by Melissa A. Kenney, a AAAS Science and Technology Policy Fellow hosted by the NOAA Climate Program Office and Assistant Research Scientist at Johns Hopkins University; Robert Chen, the Director and Senior Research Scientist at Columbia University's Center for International Earth Science Information Network (CIESIN) and Manager of the NASA Socioeconomic Data and Applications Center

(SEDAC); and Jim Smoot, manager of the Earth Science Office at the NASA Marshall Space Flight Center. A white paper was written to help calibrate thinking, frame key issues for the workshop, and lay the foundation for some of the significant elements of the NCA effort. [The White Paper has been revised after the workshop to reflect participants' comments and suggestions and is included in this workshop report.]

During the opening session, Kathy Jacobs, Assistant Director of Climate Adaptation and Assessment at the White House Office of Science and Technology Policy, remarked that one of the major efforts moving forward in the NCA is the development of indicators. The goal of this indicator framework is to identify a small number of policy-relevant, integrated indicators, designed to provide a consistent, objective and transparent overview of major trends and variations in climate impacts and our ability to respond. Such a system should include metrics; assess progress of adaptation and mitigation; and to the extent possible integrate physical, social and ecological components. Lawrence Friedl, Director of NASA's Applied Sciences Program, gave the welcoming remarks to the workshop participants. To frame the effort charged to the workshop participants, he quoted Meriwether Lewis on his thirty-first birthday:

"I reflected that I had as yet done but little, very little indeed, to further the happiness of the human race, or to advance the information of the succeeding generation."

He noted that future generations could not inform us of what priorities they had; therefore, we have to be the trustees of the future. In that regard, we must develop and implement indicators that help inform the public and decision makers about past and projected climate change impacts, opportunities, vulnerabilities, and adaptation over time.

The workshop participants were asked to provide input on a number of topics, including (1) categories of societal indicators for the NCA; (2) alternative approaches to constructing indicators and the better approaches for NCA to consider; (3) specific requirements and criteria for implementing the indicators; and (4) sources of data for and creators of such indicators. Socioeconomic indicators could include demographic, cultural, behavioral, economic, public health, and policy components relevant to impacts, vulnerabilities, and adaptation to climate change as well as both proactive and reactive responses to climate change. Participants were given explicit instructions that consensus advice was not being sought by the workshop organizers or NCA staff.

What follows is a summary of the workshop's presentations, breakout sessions, and discussions. The statements in the following sections do not represent consensus of all participants, but are general themes that emerged from presentations and individual comments regarding societal indicators during the workshop, as observed by the authors of this report and other rapporteurs and participants. See the accompanying appendices and sections for the White Paper, societal indicators inventory, agenda, list of members of the steering committee, and societal indicators bibliography.

2 INDICATOR SYSTEMS - PANEL PRESENTATIONS

The first panel discussed societal indicators for the NCA. The panelists were Tom Wilbanks, Pat Gober, Mike McGeehin, Ben Campbell, Gemma Cranston, and Radley Horton. The panelists each provided a 10-minute informal presentation that summarized indicator systems that they have developed (or contributed to), the lessons learned from developing such indicators, and thoughts for the NCA in developing an indicator framework that includes societal indicators.

2.1 Tom Wilbanks - Oak Ridge National Laboratory
Tom Wilbanks opened by pointing out that there is a rich tradition of work done on social indicators – health, education, security, living conditions, and others. He noted that indicators need to focus on vulnerability (exposure to threats, sensitivity to threats, and coping capacity) as well as resilience. For the latter, he suggested that this is not a case of whether the conditions are good, but whether the social dynamics are good. For these reasons, we need to develop composite indicators instead of indicators focused on one variable (for definitions of different indicator approaches, such as "composite," please see the White Paper). Wilbanks noted that a critical deficiency in developing societal indicators is that no time series data exist for many of the important social or economic indicators. If time series data are available, they exist in crude form (e.g., every 10 years). He made several references to the National

Academies report entitled "Our Common Journey: A Transition toward Sustainability" (1999). This report describes how we can get to a sustainable world in 50 years and includes an entire chapter on sustainability indicators. Wilbanks also noted that in the last decade, there has been much interest in connecting societal indicators with what can be observed from Earth Observation from space. This has led to a number of workshops sponsored by the National Academies and the U.S. Group on Earth Observations (USGEO).

Wilbanks indicated that there has been significant interest in a report by the National Academies on "Monitoring Climate Change Impacts: Metrics at the Intersection of Human and Earth Systems" (NRC, 2010), particularly by the intelligence community. He alluded to the concept that for developing societal indicators, it is important to ask what questions you want answered before you start working with the data available now. He suggested the need for caution because the "hunger" for indicators leads to questionable practices: this underscores the importance of validation of indicators before they are implemented. This also leads to the question of how one would validate societal indicators given the lack of time series data. There is not one set of indicators that will be equally good for all purposes. The existing knowledge base does not support what we want to know, and because of this, developing a knowledge base will require new research and possibly new data systems. On the other hand, Wilbanks pointed out that indicators of vulnerability and resilience need to be developed even if the knowledge base is not yet developed. Climate-sensitive health indicators and land use indicators need to be identified: these should in many cases not focus on current conditions, but characterize rates of change of these conditions. Moreover, indicators ought to identify society's response to risk of extreme weather events – not just responses to climate change. Society also needs to have some idea of changes in resource requirements to respond to risks; e.g., recharging water tables in water scarce areas.

2.2 Patricia Gober - Arizona State University
Patricia Gober reported on lessons learned in the six-year National Science Foundation-funded study on "Decision Center for Desert City" that is focused on determining how climate science products can be turned into tools useful to decision makers and the public. The lessons learned here relate to the

science of knowledge translation, i.e., how scientific tools are produced, communicated to, and used by decision makers and the public at large. She noted that five years ago the president of Arizona State University undertook the task of creating an indicator system for the Greater Phoenix area. Their goal was to produce sustainability indicators - those factors that connected the human to the physical system to identify trends in these human-coupled natural systems. Are we making progress toward sustainability goals? Are we approaching critical thresholds when policy decisions need to be made and implemented? They developed four sets of indicators: air quality, water quality, the urban heat island, and energy. One aspect was the charting of local temperature over time, with the data downloadable by any citizen. Other measures included cooling degree days and the extent to which the urban heat island affected energy consumption over time. They also looked at how energy consumption is linked to electrical power use; average low temperatures; number of days with a low temperature of 90°F or more; etc., as a measure of relative consumption of renewable sources.

Gober identified the lessons learned from this activity. First, citizen participants need an historical context and expert interpretation to make sense of the information transition from the data that is provided to what is meaningful to people on the ground; for example, the number of people living within walking distance to public transportation as an indicator. Second, there was considerably more interest in population and income variables than in sustainability indicators; e.g., the link between the urban heat island and potential population growth and between air quality and water quality and energy. Third, the credibility of data was highly important. The project used water quality data from the U.S. Environmental Protection Agency (EPA). One community strongly maintained that the approach being used by the EPA was wrong; it threatened to take legal action unless the data of concern were removed. Additionally, Gober noted that the project created an interactive tool that permitted people to "play out" the consequences of various policy decisions and to the future sustainability of Phoenix. This allows people to alter the indicators using "what if" scenarios to assess the implications of decisions. She reported that early engagement with stakeholders was crucial, acknowledging that the project's failure to pay

sufficient attention to this in its early phases created considerable problems. She closed by affirming that people are "intensely interested in their communities."

2.3 Michael McGeehin - RTI International

Michael McGeehin said that the new NCA process, especially the continuing assessment approach, is exciting and innovative. He described his experience at the Centers for Disease Control and Prevention (CDC) as it relates to developing societal indicators. He had the task of creating the National Environmental Public Health Tracking Network (EPHTN), which is a Web-based system that brings together data from disparate sources, meets rigorous standards, and is updated regularly. The EPHTN was developed for state health departments, local agencies, elected representatives at the federal level, and the public. McGeehin noted that there are tremendous similarities between what he did at the EPHTN and what we are trying to do with NCA societal indicators. He emphasized that continuing attention needs to be paid to the source and quality of the data, which reflect on the reputation of the various agencies or sources that provide these data. This is indeed an onerous task. It was, he said, very expensive to gather public health data; he did not believe additional funds to do so would be forthcoming from Congress. That being the case, it was necessary to make use of data sources that currently exist. The CDC put together a work group of people for the EPHTN who had vast experience in dealing with health surveillance data and climate change, and this group did come up with a suite of indicators on climate change and health. However, as he noted, linking health data to climate change was difficult. What was needed was a well-accepted epidemiological approach that tied to climate change. The EPHTN working group looked at the epidemiological scientific literature and assessed what can be associated with ecological attributes or changes in weather. This assessment of public health data was a huge effort.

McGeehin noted that the spread of West Nile disease in the United States had been first mapped geographically; then local weather conditions were added as a second variable for linkage with climate. In general, associating public health with vulnerability, mitigation, adaptation and policy indicators can be very problematic. The main problem in the health arena is that health indicators are simplistic or they cost millions of dollars to create and implement. Similar issues may ensue for developing a suite of societal indicators and climate change. McGeehin suggested that vulnerability indicators will include some that are "generic" (e.g., the aged, the young, or the flood prone). He thought that NCA indicators will probably cross sectors most easily because we have health data that are linked to other sectors, and they may be used as a suite of indicators. He ended by saying that the public always responds to health issues, but we must have high quality data that resonates with the public and elected representatives.

2.4 Ben Campbell - Millennium Challenge Corporation

Ben Campbell reported that the Millennium Challenge Corporation (MCC) was a relatively new foreign aid agency, established in 2004, and charged with rewarding the "best actors" – that is, to invest in those countries that govern effectively, encourage economic freedom, and invest in their own people. MCC measures on an annual basis whether countries are performing and encourages them to compete amongst themselves. MCC had issued an open request for proposals to help establish metrics for qualification and performance. Over time, 17 indicators have been selected and developed. To receive aid through the MCC, countries need to be at or above the median in three categories: ruling justly, investing in people, and encouraging freedom. Additionally, they have to meet the median or above for at least half of the 17 indicators. Furthermore, they need to have relatively low inflation and to demonstrate action being taken against corruption.

Campbell noted that all MCC indicators are peer reviewed. They were independently devised so that the MCC does not appear biased. Data need to be rigorously gathered, consistent, and publicly available, and have broad country coverage and comparability across countries. The program's overall goal is poverty reduction through economic growth. The indicators had to be useable as a basis for action by the governments in question; further, they needed to be annually measurable. He noted that each year the agency has gone through considerable angst as to whether the indicators used were appropriate to the task. He noted that consistency of data across countries and across time is often difficult. Also, they do not control methodologies (i.e., changing methodologies) because this causes real problems for countries. Consistency in data can be a problem because some countries are not known for a consistent approach

to data collection. These indicators are consolidated on a "score card" that shows how a country ranks on each indicator. Campbell reported that many countries placed considerable importance on the rankings they received. Because each country is competing against its cohorts, it is possible for a country to rise or fall in rank based on no action of its own. When a country reports a change in a given indicator, it is important for MCC to understand why the change has occurred. He concluded with the advice that for societal indicators we must look at what data have changed for a particular indicator and we need to know why the data have changed and need to look into this. We need to know what the indicators are telling us and if they are truly measuring what we think they are measuring (e.g., is an indicator which is intended to measure good environmental management actually doing so, or is it measuring other intervening factors). Campbell also noted that everything they do is on their Web site so that people can see where problems exist. There is also transparency in the approach, interpretation, and limitations of the data.

2.5 Gemma Cranston - Global Footprint Network

Gemma Cranston described her organization, the Global Footprint Network, as engaged in assessing the availability of natural resources and demand upon such natural capital worldwide. The Global Footprint Network base their assessments on the "Ecological Footprint" concept which is a natural resources accounting system. They are looking at the amount of biologically productive land and sea area an individual, a region, or all of humanity requires to produce the resources it consumes and absorb the carbon dioxide that is emitted compared to how much land and sea area is available in any given year. This allows for demand versus supply assessments to be made; i.e., compare total consumption with total availability to get a demand balance. Clearly demand is currently out-distancing supply. She noted that six land use types are assessed: carbon footprint, built-up land, forest, cropland, and grazing and fishery lands. The Global Footprint Network is able to make statements about historical trends in Ecological Footprints from 1961 to 2007 for some 200 countries. This provides the ability to compare the capacity of total biocapacity and the Ecological Footprint through time.

According to Cranston, the world is currently using about 50 percent more natural resources than are being created. She presented two world maps

– one with data from 1961, the other from 2007 – that show massive change. There is a dramatic shift toward a "biocapacity deficit" in this period and there is a need to identify key factors that contribute to this deficit. She noted that while many African countries have a "low Footprint" in terms of resource use, this is tied to a low-level of human development. Most commonly, countries that have attempted to improve their human development have done so at the expense of increasing their Ecological Footprint. She presented a series of charts on Ecological Footprint by economic sector: the largest deficit is in "transportation" and the largest contributor to that deficit is carbon-based fuels. In summary, the planet has limits. One needs to know the biocapacity that is available and the amount that is being used. "Blindness," she said, "costs lives and opportunities." The planet has limited natural resources and more and more people are demanding more each year. Thus, without an understanding of the Ecological Footprint, there will be more detrimental effects on people and economic systems.

2.6 Radley Horton - Columbia University and NASA Goddard Institute for Space Studies

Horton is working on comparing and contrasting two approaches to indicators related to urban areas and climate change. The first is the Flexible Adaptation Pathway, which was developed on the premise that great uncertainties always exist. This has been emphasized by New York City, where each city agency had been charged with thinking about its key mandates with respect to climate components, tracking adaptation, strategies, and vulnerabilities. His organization has been bringing these various components together; this requires standardization of data sets and getting agencies to share and communicate with each other. He believes there is a need to move beyond infrastructure issues to include ecological issues. Horton said that New York City has challenges in projecting the magnitude of climate events for the region (northeastern U.S.) such as sea level rise and ice sheet melting. Moreover, New York City is concerned about elements that will result from climate vulnerability; for example, how are populations going to change throughout the northeast as a result of climate change? Horton's research group is working with NASA on a second approach to identify the vulnerabilities of each NASA center via the Climate Change Adaptation Science (CASI) program to assess climate change vulnerabilities and environmental assets at each center. At Kennedy Space Center (KSC) for example, data have been accumulated on key variables; e.g., days above 90°F and days above 95°F, which influence worker safety. He noted other climate-related factors, such as the number of launch cancellations due to unfavorable weather. He argues that NASA needs to look at weather and adaptation in a broader context; e.g., to what extent will changes in global ice sheets impact weather at KSC, and how will changing wealth influence the use of air conditioning and thereby impact electricity use?

3 GOALS FOR THE NCA INDICATORS

During the breakout sessions, workshop participants were asked to comment on the following goals for the NCA indicators, as stated in the White Paper

1) provide meaningful, authoritative climate-relevant measures about the status, rates, and trends of key physical, ecological, and societal variables and values to inform decisions on management, research, and education at regional to national scales;
2) identify climate-related conditions and impacts to help develop effective mitigation and adaptation measures and reduce costs of management; and
3) document and communicate the climate-driven dynamic nature and condition of Earth's systems and societies, and provide a coordinated benchmark for all regions and sectors.

The following comments on the goals were made by individual participants during the workshop and are summarized below. These comments do not necessarily reflect the views of all participants.

3.1 Comments on goal #1
Focus on informing decisions at multiple scales. This goal forces the question of whether the same indicators suffice for decision makers and the general public. The stated goal should be mindful of decision-making by whom and of the need to support decision-making for multiple audiences (e.g., government, business, and NGOs). It should also help them understand what is happening (baselines, impacts, causes) by tracking trends, variability, and extremes. It is important to also include causal factors within this goal.

Engage stakeholders in both information and development of indicators when working toward this goal. Even though indicators are not usually tailored to specific industries, the NCA should engage with stakeholders to understand how indicators are used and can be made relevant to them. By focusing on the major trends and common understandings, different groups with different needs should be able to be on the same page.

3.2 Comments on goal #2

Explain whether the measures being developed are direct or proxy. If using a proxy, communicating the link is a very important part of the process.

Include risk management. As there is no direct reference to risk in the goals, risk management needs to be incorporated into discussions of how to "reduce costs of management." Risk may appear to differ at individual, community, national, or global scales. If so, appropriate scale-based insights may be valuable.

Be direct about how indicators will address mitigation and adaptation. What can indicators tell us about mitigation and what is the NCA assessing? Is this about planning adaptation and mitigation responses that can be tracked over time and what happens after implementation?

3.3 Comments on goal #3

Clarify the third goal. Clauses should be separated out; for example, the first clause of this goal seems to repeat the first goal. It is unclear what is meant by "benchmark"; is this a log-book for keeping a record of the data or is it supposed to mean the basis for evaluating the future? Furthermore, it is unclear if the "coordinated benchmark" is still in the development stage or already in place.

Replace the term "benchmark" with "framework." Because the goals are very broadly framed, they might not be as appropriate as regions and specific problems. However, someone with specific needs might want to pull from the indicators to help answer questions. Providing a coordinated "framework" would help with communication and customization at the regional and sectoral levels.

3.4 Suggested additional goals

Indicators should be used to talk about the future. Projections and scenarios can be used for some of the indicators to play out different futures to inform

decision-making. It is important to keep in mind that many people might not care about the indicators themselves, but rather are interested in what it means in terms of future projections for their area. The projections and scenarios need to be clear, transparent, and credible in process and uncertainties. Adding future projections into the goals should be kept simple to just one or two phrases. However, reporting on observations and not just projections was also seen as being powerful. Additionally, the information gaps to be filled by the indicators need to be prioritized even if current knowledge is poor.

Education should be considered an important goal. The NCA indicator effort could help educate the public. The indicator system presents an exciting opportunity for climate education using the latest communication principles and tools, such as the use of social media.

How the data will be used and the audience targeted should be added as a goal. For example, if the purpose of indicators is to promote autonomous adaptation (people adapt because they see signals around them), then this is a different goal than planned adaptation (what government does through programs and policies). Goals should be framed to develop and support policy decision-making to inform multiple levels and audiences (e.g., local government and business).

3.5 General comments on the goals

These goals are better suited for the long-term, ongoing process. The NCA needs to decide on building infrastructure for the long-term versus choosing some simple indicators that could be measured in time for the 2013 report. The NCA should be cognizant of how the goals fit into the two different timeframes. One suggestion would be to start simple and build into more complex indicators as needed.

The goals should be directed toward actually measuring adaptation and vulnerability. Currently, the goals are too impact-oriented and should instead be operationalized to clearly include vulnerability and adaptation. They should be framed to focus on positive actions that can be used to measure adaptation and vulnerability.

The purpose needs to be clearly stated. What do we want indicators to tell us down the road? Is the purpose to react to what is happening or to actually affect changes that might alter the course of events? What is the impact of the information? The goals should capture the rationale for developing indicators in a compelling way that allows them to evolve over time with strong purpose and direction.

There needs to be a framework for understanding "signal-to-noise". The indicator needs to be a clear signal of what it is measuring and should be evaluated to ascertain whether other intervening factors might be significantly influencing the indicator outside of climate trends (i.e., is there an alternative explanation for what the indicator is telling us?). This is highly important for societal indicators, and could be used by policymakers. For example, it is important to place the unusually large number of severe tornadoes and tornado deaths this year in context, such as increased hazard related to climate change and other factors as opposed to being only a meteorological phenomenon.

Indicators should inform decision-making by empowering people with actionable information. The NCA could use the indicators to provide an empowering message; the metrics could be framed to tie mitigation strategies to social benefits (e.g., health). The indicator system needs to work in parallel with levels of decisions (e.g., local, state, regional, or national). If the desired outcomes for the indicators are to get people to pay attention, to believe the information, and use them in decision-making, a process needs to be designed to accomplish these objectives.

Indicators should link to issues that are meaningful to society. Consideration should be taken on how to reach out locally to inform and incentivize people to pay more attention to climate change. The focus should be on finding things that people relate to and things that people can do something about, rather than things that are in the political realm. This includes being very careful about word choices. For example, be careful about using the word "forecasts" versus "projection" because of associations with weather to different populations.

A parsimonious way forward should be adopted. It needs to be made clear that indicators are representative, not necessarily comprehensive. However, indicators should be founded on good science that embodies confidence (statistical) and repeatability. To avoid being too compartmentalized, because climate change impacts people and places in different ways, take interdisciplinary approaches to developing indicators, but be careful about how to combine different indicator approaches.

4 AUDIENCE FOR THE NCA INDICATORS

Workshop participants were asked to provide feedback on the White Paper's statement that a "primary audience for the NCA indicators is certainly the collection of present and future legislative and executive branch leaders at federal and state levels. A second major audience is the general public in the U.S., specifically the interested and informed public". Participants also discussed other potential user groups and the connection between the user groups' needs and indicator goals.

4.1 Comments on the NCA indicators audience

Clearly articulate the audience and the goals. The most important questions to ask first when developing indicators are: what is the purpose of having indicators, how they are used, and who should they serve? The audience drives how indicators are approached; therefore, who the audience and NCA users are, who is making decisions, who will sustain the indicators over time and what the NCA is trying to accomplish needs to be identified and clarified from the beginning. It is imperative to engage the target audiences and consumers of indicator information (e.g., public and private sector decision makers) up front in validating and refining key questions, in defining indicators, and in understanding how indicators are used, as the same indicators might not be important to all people.

The framework should be flexible, customizable and serve multiple audiences. Indicators must be relevant to multiple audiences (multi-purpose), but do not have to be all things to all people. Trying to build some degree of common understanding is incredibly important. It is more practical to have a set of indicators that might be common to different people than to come up with different indicators for different groups. However, there is a need to prioritize key audiences. The NCA could examine approaches taken by states and other groups who have come up with indicators to help determine the audience. This could be part of a process for stakeholder involvement in which audiences are asked what they care about. The NCA needs to address how to make indicators usable and accessible by the general public and decision makers. This means contemplating if the same set of indicators could be used for both decision makers and the general public or if these two audiences have fundamentally different indicator requirements. Although the stated primary and secondary audiences are policymakers (federal and state) and the interested public respectively, other audiences could include the scientific/technical community, general public, managers, and regulators (see Box 1 for additional audiences and aspects of suggested audiences to consider).

5 SCOPE FOR THE NCA INDICATORS

Participants provided comments on the White Paper's suggested scope for the NCA indicators, focusing the discussion around the process, spatial and time scales, and communication.

5.1 Process

Establish the NCA as a process, not just a product. It is better to have two-way process where the science and user needs are linked and feed into each other. This will create indicators that are grounded in the best science and social science and are communicated to non-scientists to respond to stated stakeholder needs. To ensure user engagement, it is essential to bring stakeholders into the indicator process from the beginning. One method to engage the stakeholders would be to present the stakeholders with case studies or pilot studies that present potential indicators and actual information to move out of the theoretical and conceptual to the practical. The process for indicator development may be just as valuable as the outcome of the indicators themselves.

Start with things that are easy first. Indicators must have value, be feasible, be simple, and be translatable to have significance in people's daily lives. It is effective to display data visually for people to understand and translate for their use. Focus on existing indicators that are already being collected and maintained by public agencies or the private sector that help provide answers to the key NCA indicator questions and goals. Build on these indicators by focusing on combining different data sets and values to build the indicator infrastructure, while maintaining relevance and credibility. The scale of indicators should be matched to the scale of decision-making (e.g., federal, regional, state, or local levels).

Select indicators that would empower people with actionable information and speak to societal impacts and benefits. This includes database management allowing users flexibility. The users

Box 1. Audiences for the NCA Indicators Suggested by Workshop Participants.

- Federal/State/Local Government (examples of suggestions)
  - Metropolitan, county, and municipal governments; zoning boards; etc.
  - Chambers of Commerce
- Interested public
- General public
- Science/technical community
- Managers
- Regulators
- Trade and Professional associations
  - American Medical Association, American Insurance Association, Air Transport Association, Cargo Association, etc.
- Tribal communities
- Private sector
  - Business, industry, energy, transportation, insurance, agriculture, etc.
- Regulators
  - Have needs for indicators that are a little different than other decision makers
- Non-profit and Non-governmental organizations
- Defense/Intelligence community
- People who do mitigation and policy work
- Others not typically included, such as emergency planners and social scientists

should be able to decide whether to apply an indicator to societal, economic, or environmental issues and not be constrained in how they apply the indicators.

Create an ongoing capacity for the indicator framework. Think about what we can have done by 2013 and then add more indicators for the ongoing process, avoiding a "one-size-fits-all" approach, while building gradually as more comprehensive methodology and frameworks are developed. The indicators should continue to evolve and be iteratively evaluated to assess their effectiveness and to determine whether they are delivering the information needed, rolling them out in a phased manner. Maintaining indicators over a long period of time is necessary to see trends and evaluate actions, but it is expensive and requires commitment by those creating, maintaining, and updating the indicators. The NCA needs to think about the expertise needed on an ongoing basis to support these indicators.

Focus on the critical links between societal, ecological and physical indicators. Much of the research has been done within separate sectors of systems. However, indicators could be used to improve understanding of how human and natural systems are intertwined. It is critical to build these linkages, but it will not be easy. For example, greenhouse gas emissions are a physical indicator in one sense, but are also indicative of the organization of the economy, of choices being made and not made, and of policy concerns. Ecosystem services could be used as a starting point to link societal, ecological, and physical indicators and would allow for translating indicators to human benefits.

5.2 Spatial and time scales

Develop regionally- or sectorally-relevant indicators, organized within nationally-consistent categories. The NCA could develop a consistent national framework with a small list of more general, common indicators that could then be disaggregated to sectoral-, regional-, and local-scales (e.g., unemployment rate). Taking a regional or sectoral approach lets people decide what their biggest issues and problems are within a region or sector and begins to address the current lack of a general, coordinated signal. For example, NOAA's U.S. Climate Extremes Index is being downscaled to express the indicators regionally. However, there is a need to be careful about how indicators are being normalized (i.e., put into a common unit or utility function). It is also essential to keep in mind that the public is not homogenous, so using fewer indicators will be challenging. Additionally, it is important to consider the international context and indicators that link to what the international audience cares about.

Translate indicators to local contexts. There are some indicators that could be used across the Nation, but the majority of indicators will be context specific. Therefore, indicators should be able to be, to the extent possible, disaggregated down to the local level. Contextualizing indicators for local geography, ecology, or culture will help stakeholders better understand the situation and, in turn, is a way to change national consciousness.

Local communities must be involved. Integrating observations and perceptions at the local level could help assess relevance and ownership, with indicators being used to feed into local decisions.

For example, indigenous communities in Alaska have documented how ecological changes have impacted their communities, which has been critical to the government engaging with them to figure out adaptation solutions. Indicators could be aimed at helping communities prepare and respond.

Be aware of time scales in relation to investments. Some indicators might have forty- to sixty-year time scales, some have decadal scales, some have annual scales, and some have monthly or even daily time scales. Also, some sectors have very different time scales (e.g., agriculture tends to have short time scales, but energy sector has longer time scales). The variety of time scales needs to be considered when considering investments made in the long-term that are irreversible.

5.3 Communication

Engage people with what they care about and create indicators that are relevant to them. For example, PlaNYC tied the indicators they used to things people care about without mentioning climate change. It reached people through issues that matter to them. Another example is the EPA climate change indicators, a set of indicators that include pictures that link to real impacts and allow the audience to see a broad range of issues that gives them a sense of the direction of trend. Having customizable indicators and a participatory data component may be a great function for the NCA to make it useful to the general public. For example, a national drought/water management study engaged people about what was important to them by allowing them to construct their own view of the situation and what needed to be done, which highlighted the potential of the effectiveness of a shared vision planning model. An inventory of deployed societal indicators is included in Parts 3-5.

There needs to be a balance between direct link to climate and what people care about.
The primary focus of societal indicators should be on aspects of life that people are emotionally concerned about and are significant to them (even if indirectly affected by climate), such as indicators related to economic conditions, human welfare, displacement, health impacts, etc. Therefore, we need to not only think about the tie to climate, but also how much people care about the outcome (e.g., childhood asthma). This could involve taking a modular approach to indicators. An indicator can be the tip of the iceberg and still be useful; it can be

one indicator of broader impacts.

Lead with the questions, not the indicators. Engage with relevant stakeholder groups to ask them what questions indicators should try to inform (e.g., how is public safety being affected by increased climate variability?). Identify a set of key questions to be addressed, validate them with stakeholder groups, and then develop or use existing indicators to address these questions. Starting with the questions will help inform whether the information provided is useful and relevant to inform decision-making.

Create indicators that provide information that can empower proactive decisions. Create indicators that inform users and motivate them to act given the information. It is important to present both impacts and opportunities as well as vulnerabilities and resiliencies because one is not the inverse of the other and presenting both the positive and negative aspects of climate change allows one to look forward to consider a range of alternatives given a realistic depiction of what has occurred and what is projected to occur. The use of case studies as a communication tool may be useful to present potential indicators and actual information to move beyond the conceptual realm and into the practical.

Use the indicators to involve new stakeholders and expand the scope of the NCA. One possibility is to have stakeholders provide input through a wiki approach to populate indicator topics. The process needs to be open, transparent, and subject to peer review. Stakeholders, locally relevant information, and citizen science could be included. There needs to be a framework for communicating and customizing the indicators for regions and sectors, as well as a coordinated effort to get stakeholder input in the development of indicators that includes an iterative process. The NCA needs to seek and be receptive to stakeholder feedback.

Use sophisticated, current, and engaging tools to reach broad audiences. There is an emerging industry of communication and engagement technology, especially in gaming and risk communication, that has relational databases similar to climate indicators that could be used to reach broad audiences, including interacting with K-12 and higher education. Be aware of the complexity, necessity, and science of communication, including how to engage people with different practices and cultural and linguistic frameworks that accounts for all people living in the United States. The NCA could consider institutional models for government-

stakeholder interaction that would lead to a set of indicators seen as useful now, but could also be modified later. One possibility would be to use logic models to determine how indicators interact with decision-making and expectations of affecting decision-making.

6 "MUST-HAVE" TOPICAL SOCIETAL CATEGORIES FOR THE NCA INDICATORS

Workshop participants discussed topical categories they believe must be included in the indicator system, either directly or indirectly. Participants agreed that while the Global Change Research Act's (GCRA) sectors are a starting point, much has changed over the past thirty years and therefore, GCRA sectors should be used at a minimum while considering expanding into other categories. Participants suggested the following societal categories as candidates to be included in the NCA indicators: health, population dynamics, equity and justice, community capacity, cultural impacts, economy, institutions/governance, national security, thresholds/tipping points, physical/natural, and resource supply. It was suggested that although these are potential climate-related indicators, there may be other intervening factors, which in some cases may be more important than climate (e.g., is higher mortality in hurricanes an indication of intensity of storms or of a demographic shift of more people living in coastal areas?). Thus, as stated previously, before any indicator is adopted, it is an important test of the validity and policy significance of the indicators. Each category is outlined in more detail below with possible indicator topics and data (Table 1).

6.1 Important considerations when developing indicators
Address the categories in some capacity through the chosen indicator system. This includes considering if these categories include adaptation and vulnerability aspects and considering topics such as urban and the international context.

Tell a story with multiple indicators to periodically highlight an area of importance. Include in the story what is actually being done, vulnerabilities, short- and long-term trends (including historical trends and lessons learned), projections, and interaction of stressors (e.g., water implications for agriculture stress), tipping points/weakest link in the system (combination of social, ecological, physical systems). Help identify priorities through these tipping points so people can identify places where action can be taken. Allow for a greater understanding of the interaction of these indicators, and the factors they are trying to measure (e.g., all of these indicators are moving in complex ways and interacting with each other).

Design a framework rather than a static system. Consider how to most effectively frame the indicators. Security is a good framer (e.g., water security). Another way of framing is to consider what we want to be tracking twenty years from now with this indicator system (e.g., what are indicators

Table 1. Categories and Potential Indicators (Note: At this point, the indicators and broad categories have not been linked to specific climate-relevant questions that could be addressed by an indicator. Future work should link such indicators to the NCA indicator questions and goals.)

Category	Indicator
Health & Safety	Weather-related mortality (e.g., heat, floods, and wind) Weather related illness (e.g., hospital admissions for heat stress or heat stroke) Vectors (which ones and how they change) Chronic health conditions (e.g., asthma) Health in vulnerable populations Healthcare access Healthcare systems Birth rate Mental illness Subjective well-being (e.g., "happiness"; includes social cohesion, civil society, and occupation structure) Quality of life Infectious disease risk/geographic extent (e.g., malaria; potential and actual) Number of people experiencing heat waves multiplied by the number of days Air quality Safety (e.g., crime rates)
Population Dynamics	Socioeconomic dynamics, raw population, demographics, race and ethnicity (as associated data set) Human well-being (composite indicator – sense of place; stability, feel like living in risky environment; health; cost of living; community habitability; effects on recreational opportunities; how much time spent outside) Vulnerability (populations, regions; exposure, sensitivity, and adaptability; socioeconomic vulnerability; population sensitivity; elderly and family structure) Settlement and movement (displacement, migration, location of populations; population density; population change; population distribution; e.g., natural amenities scale, which includes climate, topography, and access to water to understand depopulation in rural communities) Social network mapping Persons in 100- and 500-year floodplains and coastal storm surge zones "Special needs" populations in those zones Social capital, connectivity and networks (includes population learning/literacy/attitudes; knowledge, action, and practice) Societal awareness of climate change (understanding; communication; education; attitudes; climate literacy) Behavioral shifts in transportation (alternative transportation; e.g., indicator by Department of Transportation about how long we sit in traffic and congestion patterns, which lends to greenhouse gas emissions and affects quality of life) Intergenerational Social disruption (e.g., communities affected by hurricanes; long-term consequences)
Equity & Justice	Socioeconomic inequalities (spatial/place-based; who can move; social networks; access to services, infrastructure, institutions; middle class crunch; ability to cope post-disaster) Environmental justice (exposure, vulnerability, resiliency; subsistence practices; housing; e.g., subsistence practices that are being modified, impacts on livelihoods)
Community Capacity	Risk Stress Community habitability (displacement of entire populations from an area) Response capacity (potential vs. action) What is actually being done for mitigation and adaptation (e.g., money spent on flood proofing and other hazard mitigation spending) Transformational adaptation (anticipatory)
Cultural Impacts	Aesthetic environment (e.g., color lost in leaves in Smoky Mountains) Cultural richness of communities (i.e., Richard Florida's "Creative Class") Impacts on cultural practices Cultural processes Cultural icons (e.g., maple tree) Cultural identity Human social systems, ways of life

Category	Indicator
Economy	Climate investment (resiliency, adaptation and mitigation) Risk of investments Economic assets at risk Direct/indirect economic loss/gains (e.g., increased production in warmer/wetter areas) Infrastructure (e.g., communities relocating in Alaska are living in public health crisis because government no longer investing in their infrastructure) Lives Climate risk reduction and costs (cost-benefit) Economic diversity Tourism Agriculture Forestry Employment/unemployment (in certain fields) Energy Insurance (e.g., property losses from extreme weather events) Change in when people work (e.g., as a result of temperature, heat index, precipitation, power outages) Lost work productivity Overtime work
Institutions / Governance	Institutional (learning) capacity Flexibility and adaptive management (how flexible is infrastructure and institutions) Institutional coordination – governance and leadership Government structure, changes in governance Civil society Tax base Costs/responsibilities Emergency Medical Technicians/healthcare workers Emergency preparedness plans Early warning systems Effectiveness of communications (e.g., early warning systems) Index focused on knowledge systems and innovation (ways to monitor progress, advancement, conditions for innovation) Preparedness Response capacity (potential vs. action) Insurance and reinsurance Intergovernmental issues What is actually being done for mitigation and adaptation Confidence/attitudes about government (sense of security and confidence in government) Rate of emissions and what trajectory it puts us on for long-term
National Security	Security and confidence Water security Food security Energy security Housing security Resource conflict
Thresholds / Tipping points	Extreme events (probabilities; number of 100-year or "greater" events) Climate change added to other stressors (e.g., storms in coastal areas combined with sea-level rise)
Physical / Natural	Heat (air quality; ozone and particulate matter) Precipitation Urban heat island Land cover and change Human feedbacks on local and regional climate Weather extremes (catastrophic; mortality rates) Location/duration/timing/severity of precipitation, drought, flood Agriculture (food security; livestock and crop disease; fisheries/forests) Ecological feedback loops (unexpected /surprises) Coasts Vegetation migration Biological diversity Ecological health Wildland/urban interface issues (e.g., wild-land fires; also related to population migration) Increase in hazards
Resource Supply	Water (i.e., quality, quantity, availability, access, and provision) Acre/feet of water supply in reservoirs Food Energy (production; use; consumption patterns; different sources) Land resources Food security /agriculture Coastal Storms Ecosystem services

of change that society is transforming; and what we have done to change the drivers of stressors and our society's ability to cope with stressors). Another framing possibility could be a risk assessment framework (hazard, exposure, effects, risk). The framework could also include current trends in emissions and account for what is happening on mitigation at the state and federal levels. Possible approaches to start with are the precautionary principle or with physical changes that go through to impacts on people. The framework should consider how societal, physical, and ecological indicators relate to each other.

Create a library of indicators that meets a variety of needs. Indicators do not have to be inclusive, but can have broader impacts. A mechanism could be a library of resources of twenty categories, with a framework for sub-categories on which others can build. The suite of indicators must allow for local customization, include the best practices, methods, and decision models to relate to different users, and include the space for appropriate development. This could also include maintaining a "thermometer of innovation" to identify how people come up with innovative ideas to solve problems, providing an opportunity to consider what institutional capacity looks like when linked to innovation. One suggestion would be to create an indicators warehouse with a collaborative space for application development. This could also include providing guidance on when it is and is not appropriate to use an indicator. The NCA could be used as a holding ground for information and linking to various indicators. A virtual set of indicators could be established that directs users to where information can be found and/or reported.

Leverage existing efforts and lessons learned. Because of the difficulties in sustaining (maintaining and funding) indicators over the long-term, it is important to focus on using existing indicators that are already being supported by public and/ or private entities. Keep in mind previous mistakes and successes as we move forward, such as the tendency to neglect inter-generational and intra-generational populations. There are also gaps in terms of populations and communities for which the indicators are targeted; for example, indicators are not often targeted at the elderly. It is also important to look to existing influential tools and effective ways of presenting information. Some examples include the U.S. Geological Survey's report about how water is being used, for what purposes, and

how it might change; the National Integrated Drought Information System (NIDIS); and the United Kingdom's Foresight process that looks at scenarios and is influencing policy and action in the United Kingdom.

Link indicators to meaningful outcomes that speak to societal impacts and benefits. This includes positioning indicators so that they are actionable to the users. It is worth creating a core group of indicators that provides people more than just data, but translates that data into an indicator that addresses critical climate questions for the audiences, an important component for longevity.

Establish key criteria for indicators. There should be a decision matrix to select and rank indicators, with at least three key criteria to choose indicators, such as: 1) climate connection, 2) audience resonance, and 3) data quality.

Components of indicators may be the most important. Metrics that are important for an indicator may be different for different communities or regions (e.g., water is looked at differently in the eastern and western U.S.) and some environments or communities are more stressed than others with different levels of resiliency. Some framing options include the needs of future generations or stressors on communities.

7 INDICATOR PROS/CONS AND LESSONS LEARNED FOR DIFFERENT INDICATOR APPROACHES

Workshop participants were asked to consider benefits and drawbacks of different indicator approaches, as well as lessons learned to incorporate when developing the societal indicators. They were asked to draw from the indicator approaches discussed in the White Paper and to add additional approaches. The specific indicator approaches discussed include composite indicators and indices, dashboards and baskets of indicators, and "systems" or accounting indicators. Participants were also asked to comment on considerations for choosing among the three approaches.

7.1 Composite indicators and indices
Participants noted that the benefits of using composite indicators include the ability to communicate real, tangible opportunity costs, and the ability to compare, rank, and consider the big picture. For example, composite indicators

could be created to demonstrate the opportunity cost of more money spent on water investments in California to save the wine industry from climate change. However, participants also noted that there are potentially negative aspects of using composite indicators, including that 1) composite indicators are developed for a specific purpose, 2) they are less transparent, 3) they mask the role of independent factors that go into the composite, 4) they may include normative weights either explicitly or implicitly, 5) they do not explain well the causes of vulnerability and do not work well for reducing vulnerability for specific interventions, and 6) they are more difficult to communicate because of marginal changes.

7.2 Dahsboards and baskets of indicators

Participants believed the benefits of using dashboards and baskets of indicators (see White Paper for help with these concepts) are that baskets permit the inclusion of multiple types of indicators (e.g., economic, health, and well-being), are easier for communication (as long as assumptions are detailed), and allow people to obtain information on the factors that most interest them. The basket approach was suggested as being a good option for economy, well-being, and weather categories. However, some aspects of dashboards and baskets can also be negative, such as lack of clarity when individual indicators display inconsistent trends. Using a combination of composite and dashboard approaches was suggested, especially if the NCA will be Web-based.

7.3 "Systems" or accounting indicators

Participants discussed the benefits of using "systems" or accounting indicators, which help to illustrate trade-offs in common metrics and are useful for understanding organized systems. The negative aspects of this approach are that accounting indicators often do not relate to what people care about and are difficult to make actionable, everything has to be in the same units of analysis, they incorporate too many value judgments, they might not work well with scale issues, and they might not be of great interest to the general public.

7.4 Considerations for choosing an indicator approach

Credibility of information, meeting data quality standards, and being clear about assumptions being made are extremely important. This includes considering who the trusted sources of information are (e.g., opinion leaders and knowledge intermediaries) and using these people to help disseminate information and engage them in the process (e.g., work done by Texas A&M on who people trust).

Start with building blocks and aggregate as appropriate. Approaches might need to be mixed and matched depending on the goals and audience (e.g., public and decision makers). It could be useful to test approaches with audiences to see which one(s) make the most sense and/or communicate the best and what types of data would be most beneficial (qualitative versus quantitative). This engages the users up front in the process, which would also help assess the use and possible misuse of indicators by learning how people intend to use the indicators.

Explore additional approaches. One idea is to take an outcome approach, which means focusing on outcomes and then trying to understand the most significant predictors and where they are located. Another approach is to base the indicators on a few qualitative categories rather than numbers, which could be used to evaluate things like resilience. Providing people with information allows them to define problems and opportunities. Another idea is a phased approach, which starts out with establishing the current status and then call for more research to explore tipping points. Further research is needed on how different approaches might support an understanding of climate impact, as well as more research on underlying data systems.

Start with the question(s) you want answered and then select approach(es) that fit. Choose a meaningful context of indicators that allows an audience to say "so what?" This includes considering what can best be communicated to decision makers and keeping in mind what different pieces could be used for different audiences, which could then be rolled up for higher levels of decision-making, while avoiding being too policy prescriptive.

8 INPUT TO THE NATIONAL CLIMATE ASSESSMENT DEVELOPMENT AND ADVISORY COMMITTEE

Workshop participants also discussed what they would like to share with the National Climate Assessment Development and Advisory Committee (NCADAC), the federal advisory committee for the National Climate Assessment. Views expressed here are those of individuals and do not represent consensus inputs from the workshop participants.

Develop **methodology and process** to adequately articulate and communicate the indicators to decision makers and general public; this includes considering the most effective technologies and methods to gather data (e.g., climate surveys) and integrate them into a useable format. Consider methodologies that allow for a multi-scale analysis. It is essential to think about what our desired outcomes are from an indicator set. Build a flexible framework and construct the infrastructure for a sustainable delivery of indicators. Start small and build up to the long-term vision. Make sure that the indicators are feasible, repeatable, relevant, meaningful to users, and transparent. Evaluate the effectiveness of the indicators in a scientifically rigorous fashion. There should be **ongoing evaluation** of the approach and of individual indicators to ensure that the indicators are useful and relevant.

Roll up indicators into a story. Stories should be at state or sub-state level to make indicators compatible with state-level decisions and to reach out to local stakeholders, while recognizing political boundaries. Include quotations from industry leaders and include safely generated economic numbers and local, good quality data when possible. Also, use the breadth of data across agencies.

The NCA should keep to indicators that are **simple and transparent**, using experiential and observed data. The NCA should be wary of complex indicators that require value judgments, including efforts to evaluate vulnerability indicators through empirical models. Caveat the indicators properly and leave time for vetting and iterations before going public. Leverage credible, existing efforts, while choosing indicators that are flexible enough to include regional and cultural differences and reliance on ecosystem services.

Be realistic about what is **achievable** with the time and budget. This includes considering what can be done now, what can be sustained, and where the NCA can partner with entities outside of the federal government. Sufficient time is needed to think indicators through, with guidance given to federal agencies on how to continue with indicators. The NCA needs to be clear on what types of decisions will be based on or influenced by indicators, paying attention to winners and losers created when a set of indicators is selected. Indicators could be used as a way to fill in essential needs if we do not have the data we need now.

Use **multidisciplinary, local community engagement** to go from framework to concrete indicators. Public officials and business leaders need to be in it for the long haul (e.g., Boston, Charlotte, Santa Cruz, Jacksonville, and Seattle have set up long-term indicators). Engage sectors, regions, and tribal governments now in the process of developing indicators and make sure that they are in agreement and have given input to vet relevance, importance, selection of indicators, and data sources. The NCA could have listening sessions, such as tribal input sessions, to link to established sources and understand what groups track to see how a climate dimension could be added. Incorporate discussions of indicators in the NCA listening sessions (e.g., regions and sector meetings). The listening sessions could be a good venue to determine what indicators could be most useful to stakeholders on the ground. Make sure that decision makers have the necessary information by focusing on communication, education, climate literacy, and understanding. While the NCA could track knowledge of climate, it is also important to understand audiences' value judgments and behavior, including considering **how people make decisions and how information can be more effectively communicated** (e.g., Yale and George Mason climate data centers, and Columbia University's Center for Research on Environmental Decision Making (CRED)). Indicators of progress could be included, such as possibilities of tracking communities that have shifted in mentality. Our decisions can really put more or less people at risk given a changing climate.

Communication is critical. Recent studies have increasingly demonstrated that communication is as much about the messenger as the message. Therefore, the NCA should seek out a diverse group of trusted spokespeople and platforms to disseminate to effectively reach different audiences, engage stakeholders, and educate the public. The use of social media (e.g., create a Facebook site) is one popular way to reach a diverse segment of the public. Communications and presentation of information are as important as the information itself; think about graphics/presentation style for indicators from the beginning.

9 DEVELOPING INDICATORS TO ADDRESS CRITICAL CLIMATE-RELEVANT QUESTIONS

The NCA is a gateway that should link to other things that are being done and **integrate different knowledge systems**, including indigenous knowledge, and multiple information sources – both qualitative and quantitative – while assuming the rigor of peer review. It could also be used to look across potential linkages and interactions between physical, ecological, and societal components. It is important to consider if the NCA is an authoritative statement (as a report) or if it is envisioned as part of a community effort (e.g., online presence).

The goal of this breakout session was to envision societal questions for one of the climate-focused questions. The groups started by digging more deeply into the choices that people might actually make. No indicator system will fit all goals and desires; the intention of the discussion is to illuminate issues and provide input. The groups were asked not to agonize too much about making the right choice; rather, they should make a choice that people could think about.

Each breakout group focused on a different climate-focused question that could be addressed with indicators. Specifically, they considered issues of using and designing climate indicators to address the following questions:
- Are important **climate impacts** occurring or predicted to occur in the future?
- Are we **adapting successfully**?
- What are the **vulnerabilities and resiliencies** given a changing climate?
- Are we **preparing adequately** for future climate change?

All groups were asked to consider four broad questions when thinking about how climate indicators could address one of the questions above:
- How can the diverse requirements of social

indicators that may evolve in future years be *monitored* annually through a small set of indicators?
- What are appropriate *temporal and spatial scales* for assessing current and future impacts, adaptation, vulnerability and resilience, and preparedness to a changing climate?
- What *validation* strategies could be developed to provide insight into how to monitor the effects of climate?
- How could these indicators be *used by groups or individuals to broadly inform decisions*? What are their limitations?

In addition to these questions, each group had additional questions that were specific to their topic area (the specific questions are included in each of the sections below).

Because the groups were focused on thinking about indicators that would address these different topics, each group took slightly different approaches, thus there is not a unified approach to presenting the outcomes of the breakout groups, but there are a number of similar themes that arose from each of the topics.

9.1 Climate impact indicators
The climate impacts group focused broadly on the question "Are important *climate impacts* occurring or predicted to occur in the future?" To address this question, the group considered the common breakout group questions, as well as:
- Should these indicators be primarily diagnostic in nature or include predictive elements?
- Can these indicators be linked systematically with the physical and ecological indicators?
- Should the focus be on monetary impacts, mortality and morbidity, or other impact dimensions, or some mix?

One of the challenges of climate impact indicators is that they are sector dependent, so the group did not think the NCA could create a single climate impact indicator that encompassed all impacts across sectors. Additionally, when the group refers to climate impacts, it is really a discussion of impacts and opportunities because some locations or sectors will benefit from a changing climate, and it is important to capture both aspects. Because of the diversity of climate impacts, the group focused on developing indicators for specific climate impacts or sectors. The group discussed a range of topics that should broadly be considered for impact indicators, which are meant to be neither inclusive nor exhaustive, including:

- frequency and duration of extreme events,
- social cohesion,
- civil society,
- economy (including jobs and trade),
- ecosystem services,
- energy,
- geography (including demographics and population displacement),
- governance,
- natural resources,
- health, and
- infrastructure and the built environment.

The group emphasized that there are a number of considerations when developing climate impact indicators. Specifically, the best approach to develop and to present an indicator depends on the question that the indicator is meant to answer, the sector addressed by the indicator, and the available existing or new data. For example, some climate impacts could be quantified as a single number, some could be presented spatially in a map (using the map as the aggregating tool), and others would be better presented as a suite of disaggregated indicators that provide a picture of climate impacts on a particular sector.

The group discussed the broad characteristics that are important to consider in developing an indicator framework that will consider the impacts and opportunities due to a changing climate. The goals, development, and implementation of the indicators may, by necessity, be different in the short-term compared to the long-term. Additionally, some of the impacts cannot be directly measured or quantified; in those cases, using proxy measures or rigorous qualitative data approaches are essential to appropriately capture the climate impact.

Additionally, climate impacts are rarely due to a climate stressor alone; they exist in a multi-stressor context, where humans are both affected by and affecting these stressors, making the climate signature even more difficult to identify. Moreover, these climate impacts manifest themselves differently in different local and regional areas.

Characteristics of effective indicators are that they
- communicate effectively to decision makers,
- include measurable variables or metrics,

- identify data gaps, and
- include uncertainty.

Additionally, the group identified a number of factors to consider when choosing indicators such as:
- whether they are climate-driven versus climate-sensitive,
- whether there are measurable, credible data to populate the indicators,
- if nested indicators are desired, whether there is nested data, and
- if impacts should be monetized or non-monetized.

To add specificity, the group focused on two different sectors that have a large amount of data information – health and infrastructure – with potential climate impacts. For a climate health impact indicator, it makes sense to use a risk framing for the indicators; specifically, indicators could include the magnitude of risk, identify who is at risk, anticipate risk, and assess benefits. It is important to present the indicator at the temporal scale of the risk (e.g., seasonal, acute effects, and chronic effects) and consider the spatial context. In addition to describing the past and current impacts, it is also important to make projections of the potential health impacts given a changing climate. Making such projections can be incredibly difficult in part because identifying the climate signal is challenging for a number of health impacts.

A number of groups were identified that could use health-related indicators, such as elected officials, decision makers (both households and organizations), media, investors, state, local, and county health departments, emergency responders, private insurers, healthcare providers, educators, public groups, nongovernmental organizations, advocacy groups, and the scientific community. Depending on the user and their interest, needs, and level of technical sophistication, the indicators could be presented or the information could be repackaged in different ways; a Web-based deployment could help to facilitate this customization of the indicators.

The group also discussed developing indicators for infrastructure and the built environment, which clearly links to the transportation and energy sectors. Infrastructure was an interesting indicator to consider because the effects of a changing climate, such as sea level rise or increased temperature, impact a range of built structures (e.g., roads, ports, and buildings). For infrastructure, the indicators may be different if describing current conditions (diagnostic) versus future conditions (predictive). For diagnostic indicators, infrastructure indicators may consider the changes in planning maps (e.g., USDA planting region maps that showed changes in plant ranges because isotherms had shifted northward and the frequency of storm or flood events is changing such that what used to be considered a 1 in 100 year event is now more frequent), practices, and maintenance of the structures. For prognostic indicators, there may be a different indicator choice if it is informing short-term versus long-term decisions because short-term decisions may involve retrofitting current structures whereas long-term decisions are planning investments for structures that may have a 50- to 100-year lifespan. Thus, an indicator could be developed that presented the information visually (e.g., a map of infrastructure that is predicted to be inundated in the next 50 years), monetary valuation of infrastructure at risk, or quantified impacts such as wait time or number of people impacted.

The group believed that it is important to *define the question* that NCA would like to inform and answer with an indicator or suite of indicators; if the questions are not defined then indicators could be chosen that would not meet the goals and be useful to NCA users. To ensure that the indicators are useful, it is important to engage local stakeholders from the start of the indicator development process. Scientists may not necessarily be the best way to ensure quality stakeholder engagement because scientists often do not effectively communicate well with those outside their expertise. To help ensure effective communication and engagement, it would be useful to include in the workshops and other NCA activities "science translators" to help translate and communicate complex scientific information to non-scientists. Additionally, the group believed that the use of storylines or narratives that resonate on an emotional or cultural level are likeliest to be effective; further, that although messages might need to be simplified for the general public, the data behind those messages must be transparent. Finally, the group thought it is essential to engage key professional groups and societies (e.g., American Society of Civil Engineers, National Association of Home Builders, and American Water Resources Association), policy makers and media representatives in the NCA indicator development, workshops, and listening sessions.

9.2 Climate adaptation indicators

The climate adaptation group focused broadly on the question "Are we *adapting successfully*?" To address this question, the group considered the common breakout group questions, as well as

- What indicator approach could provide usable information on both planned and autonomous adaptations being implemented in different sectors and regions?
- Is it possible and important to measure both adaptation investments and adaptation success?
- Should such indicators encompass institutional, legal, economic, and technological options for adaptation, e.g., the availability of insurance, the existence of adaptation plans, investment in research, and disincentives for maladaptation?

The group believed that interactions with stakeholders were essential. Those engaged, however, would benefit from research that clearly tied investments to outcomes. The group noted, first, that investments were motivated by factors other than climate and, second, that there may not be a match between possible areas of investment and subjects of public concern. There is an opportunity cost associated with adaptation (i.e., what options were we foregoing by investing in one approach over another?). The group acknowledged that success was in the "eye of the beholder." Decisions are often enormously value-laden involving unforeseen consequences: an adaptation that one views as successful might have unconsidered negative impacts elsewhere. The group noted that practical limits exist to what may be measured. Efforts need to be communicated well to the public – perhaps, by framing them in story form.

There are challenges related to monitoring: the current data gathering system is highly decentralized; a great many variables are not being measured; much of the data input was not motivated by climate concern. The group suggested that work should begin by making use of existing reporting systems; engaging both NGOs and the private sector, and regularly undertaking the collection of data that is easily available. One model offered was that of the U.S. Census Bureau, which each decade makes a comprehensive effort that is supplemented by less extensive annual activities. The appropriate frequency with which data are captured depended on the indicator in question.

With regard to alternative indicator approaches, the group thought that a "basket of indicators" was the most commonly mentioned and the most easily communicated. The group believed that general public understanding of "composite" indicators is limited (e.g., adaptive capacity, societal learning, and indicators of surviving versus thriving). When systems or accounting indicators are used, it is difficult to illustrate causality; further, these presented a high demand for quantitative data.

The group identified a number of ways that indicators could be validated. These included turning an indicator over to a user community; tracking the expectations produced by the model against actual events, and organizing focus groups of pertinent stakeholders. The group acknowledged that indicators have their limitations. First, in many cases, the phenomenon one wishes to measure is not clearly understood; this, in turn, could undermine an indicator's credibility. Second, indicators can be misused, as their interpretation is subject to "the eye of the beholder."

The group's overarching comments were that time lags create challenges for monitoring and tracking; further, climate change does not occur in isolation from other types of changes. The group suggested that a National Census on Climate may be needed to collect data related to climate change (e.g., climate-related questions might be included in the periodic American Community Survey undertaken by the U.S. Census Bureau). In addition, the group emphasized the need for effective communication; the need for additional research to build the evidence base; a review of the existing literature to help clarify what reasonable expectations for adaptation might be and, finally, that the number of adaptation indicators be limited until the field is better understood.

9.3 Climate vulnerability and resiliency indicators

The climate vulnerability and resiliency group focused broadly on the question "What are the *vulnerabilities and resiliencies* given a changing climate?" To address this question, the group considered the common breakout group questions, as well as
- What existing vulnerability and resilience approaches could be adapted to the NCA needs, or are new approaches needed?
- Are there robust measures of vulnerability and resilience that could be incorporated into a composite measure?

The group emphasized that there are both differences and linkages between vulnerability and resiliency. Vulnerability is related to the risks or threats to a system. These threats, or stressors, occur at a range of temporal and spatial scales. Resilience is the system's ability to respond to, cope with, and recover from those stressors. Therefore, the two topics are inherently interactive and need to be evaluated together. Vulnerability assessment is based on threats; indicators assess a system's vulnerability, providing information to effectively respond to impacts and subsequently recover in the long-term.

The group identified a number of broad categories of indicators. The relevant indicators are both system-based (i.e., the strength of the physical system) and society-based (i.e., the ability of people to take appropriate action) and include elements of exposure, sensitivity, barriers, and adaptive capacity. The group identified a series of exemplary metrics focused on health and human demographic patterns (immigration and emigration) including
- frequency and severity of extreme events,
- economic status,
- social capital,
- infrastructure – including its age and cost to maintain or replace,
- knowledge and awareness,
- values and attitudes,
- available resources,
- institutional capacity, and
- mortality and morbidity.

To verify and validate the indicators, the group suggested that an indicator needs to supply the information actually needed by decision makers, be responsive to climate changes, and be regarded as valid and useful by intended audiences. This last point requires conversations with stakeholders and communities to determine how the information would be used. For example, indicators could track changing health outcomes (e.g., early warning systems implemented and changes in hospitalizations), changes in insurance practices and rates (e.g., changes in property damage/loss), economics (e.g., costs avoided or costs incurred), and lives saved or lives lost in a disaster. Additional comments intended for the NCADAC related to the long-term vision for indicators included
- there is no single number that can adequately

capture vulnerability and resiliency;

- vulnerability and resiliency assessments are inherently multi-disciplinary and thus the approaches must integrate across disciplines;
- the spatial and temporal scale relevance is critical;
- the study of how vulnerability and resilience are related is a long-term undertaking; and
- the methods by which indicators are selected and used must incorporate the ability for adaptive learning.

Relative to preparations for the NCA 2013 report, the group suggested that the priority needs are to review indicators already in use (both climate change indicators and societal indicators), assess their value, and determine to what extent they are transferable to climate-related issues. New approaches should be created only to fill identified gaps, not to duplicate any current activity. The most stressed point, however, is the need to engage the communities in which indicators would be used. As part of this engagement, efforts should be made to teach people what the indicators and the data mean. It should be borne in mind that people are more receptive to communication framed around co-benefits, resilience, and improving societies than around vulnerability. Further, efforts will be needed to clarify who benefits from a particular activity, and how.

9.4 Climate preparedness indicators

The climate preparedness group focused broadly on the question "Are we *preparing adequately* for future climate change?" To address this question, the group considered the common breakout group questions, as well as

- Are specific indicators needed and feasible to characterize the actions that federal, state, and local government and other nongovernmental stakeholders are taking or could take to improve preparedness for climate change?
- Can measures of assessment or response capacity (e.g., for natural disasters or financial disruption) be adopted to address NCA needs?
- To what degree should awareness of and education about climate change issues be taken into account?
- Do private and public sector organizations have adequate expertise and planning mechanisms needed to ameliorate climate impacts, foster effective adaptation, and address climate vulnerabilities?

This group focused on the issue of disaster preparedness, as a case study of how indicators might be developed and used to assess and address extreme events, including societal impacts, adaptation, and vulnerability. There are two key areas for thinking about potential indicators:

1) Indicators about the ability of society to provide *warnings of extreme events* that would allow those impacted or potentially impacted to better prepare and respond; and
2) Indicators of *post-disaster capacity to respond* by providing needed assistance, supporting a rapid recovery, and reconstructing damaged infrastructure, systems, and property to at least their pre-disaster condition.

Many different aspects of climate change may be relevant to society's ability to deal with extreme events and complex disasters. For example, it would be useful to have indicators on the frequency, intensity, spatial distribution, and potential changes over time of various weather- and climate-related extreme events, such as drought, floods, cyclones, heat waves, cold waves, tornadoes, and other severe storms. Climate change may also raise the likelihood of multiple stresses on critical infrastructure and response capacity; for example, in river deltas subject to flooding, siltation changes, sea level rise, and coastal storms.

The group identified a number of potential indicators related to warning capacity, such as

- Existing early warning systems: type, effectiveness, and reach;
- Awareness and education about warnings and responses;
- Extent of drills and training activities by responders;
- Status of conventional media for communicating warnings;
- Role of social media in warning; and
- Equity issues related to access to warnings (e.g., due to language, poverty, literacy, and remoteness issues).

A somewhat different type of warning capacity is the ability of the scientific community to determine when multiple "record years" reflect a trend warranting a response.

Potential categories of indicators related to response capacity discussed by the group include

- Federal facilities, response resources, expertise for dealing with climate-related disasters (for example, on the part of the Department of Homeland Security (DHS), the Federal Emergency Management Agency (FEMA), the U.S. Coast Guard, the National Guard, and others);

- The extent to which existing disaster/evacuation plans address potential climate-driven changes in hazards;
- The extent to which agency adaptation plans address disaster response needs;
- The role of public and private insurers and their ability to provide needed recovery/ reconstruction resources;
- Health system capacity;
- Power grid and other critical infrastructure;
- Vulnerability and resilience of supply chains; and
- Role of local government, NGOs, and individuals.

The group noted that operationalizing these indicators will clearly require information sharing about preparedness by the involved agencies and groups, which is a major challenge given the sensitive nature of the data. In some cases, some data are already available for related indicators and just need to be analyzed and adapted. For example, levels of disaster awareness and training in schools could serve as a surrogate for assessing wider public understanding of disaster preparedness needs.

A key issue highlighted in discussion is whether or not private and public sector organizations have adequate expertise and planning mechanisms to improve preparedness. For example, some insurance and reinsurance companies are starting to take into account potential changes in climate extremes in their actuarial tables and premium structures. Various Federal agencies have established "adaptation task forces" which are starting to address adaptation decision-making and data and information needs. Private sector companies are beginning to recognize the potential for large-scale disruptions to their supply chains due to various hazards, and some are developing risk management plans. In the area of critical infrastructure, there are important questions about whether the energy, health, and transportation sectors have the capacity to assess their vulnerabilities with respect to multiple stresses and disasters. For example, in an era of tight budgets and closing hospitals, does the health system have the ability to deal with multiple large climate-related stresses such as a flu outbreak, cold wave, and snow emergency or a heat wave and associated power outages, drought-related water

shortages, storm injuries, and infectious disease outbreaks?

Another issue discussed by the group is the appropriate temporal and spatial scales for developing preparedness indicators. Response capacity often varies by political or administrative unit or jurisdiction, whereas climate-related stresses often cross such political and administrative boundaries. For the purposes of the NCA, the group thought it might make sense initially to examine possible indicators at the national or federal level, which will be important for other scales and could serve as a model. This also enables assessment of critical infrastructure such as energy, transportation and health networks, and facilities across multiple scales.

The group recognized that current climate extremes are important tests of preparedness—sometimes society is well prepared and able to reduce or ameliorate adverse impacts, thereby preventing "disaster," but at other times the extremes overcome existing levels of preparedness and lead to adverse impacts that might have been avoided. After the disaster, event analysis is essential to understand whether indicators of preparedness accurately characterized actual levels of preparedness and if they are suitable for understanding future preparedness given potential changes in climate extremes. Engaging disaster response/management agencies and other stakeholders in developing and evaluating indicators could help improve the validity of the indicators and ideally lead to their wider use in decision-making.

10 PATH MOVING FORWARD - PANEL PRESENTATIONS

The last panel discussed the path moving forward in developing societal indicators for the NCA. The panelists were Dave Cleaves, Tom Wilbanks, Carol Kramer-LeBlanc, and Jim Buizer. Panelists each gave a 10-minute informal presentation that reflected upon the workshop discussions and provided suggestions on how the NCA might develop indicators.

10.1 Dave Cleaves – USDA Forest Service

Dave Cleaves noted that he was on the "demand side" of the indicators. He observed that much time is spent in Washington developing "great schemes" for supplying information to people "out there," whereas the demand side consisted of hundreds of decision processes that were already in motion. He urged those in the room to try to develop an intimate knowledge of the norms, needs, and expectations of the decision makers who are the users of the information. The processes in place use information on indicators, performance measures, etc., but what needs to be done is to improve knowledge of decision makers' demands, contexts values, and norms for specific indicators. Cleaves noted that much of what happens with climate change will be in response to stresses that are already being dealt with; for the past century, for example, the Forest Service has been managing a multi-stressor complex. Climate change was now being added as an additional factor. Adaptation to climate change will be a change in behavior in response to stresses that we already feel. We need to ask what the added value is of indicators for managing within the multi-stressor complex and create multiple objectives. What is needed is the creation of a climate change application that could be added to those processes already in use so that we can understand what those decision processes are. He believed that the role of indicators was sometimes understated in shaping new decisions; not only do such indicators affect current decisions, but they can identify problems not previously recognized and frame decision processes that have yet to be started. There is a well-accepted body of indicators that can be used as climate change indicators and thus, we need to link to them. He noted that whole industries were being developed around the world on the issue of sustainability. Those at this meeting, he pointed out, have the opportunity to create an adaptation industry within the sustainability concept.

Cleaves commented that risk management has become a core function of every company and organization, wherein they manage continuously for multiple risks. Sustainability has traditionally assumed some underlying tendency toward balance and the consequent idea that "things are going to stay pretty much the same." Climate change suggests that this is not the case; therefore, the question of adaptability needs additional intelligent attention. We need to bring the climate signal into sustainability using the concepts of vulnerability, sensitivity, exposure, adaptive capacity, and resilience.

10.2 Tom Wilbanks - Oak Ridge National Laboratory
Tom Wilbanks returned to a central theme: the

goals for the 2013 NCA report and the goals for the long-term sustained process that the NCA is working toward are not the same. He noted that indicators are tools not objectives in themselves. Our main challenge is to build societal indicators into a long-term assessment. He noted that he was impressed by Lawrence Friedl's comment that it was more important in the 2013 document to get the objectives right than to get the tools right. The group needed to determine what climate change risks merited the most attention. The main challenge is to build indicators that would fit into longer-term national indicators. The starting point for this discussion is to ask, "What are the ideal indicators one would wish to have in fifteen years, and that can be used to inform the Nation?" We then can develop composite indicators when we know what the indicators are. Once this is established, discussion of how to create those indicators can begin. One aspect of this discussion will be to determine what aspects of indicators can we measure with existing data and which potential indicators might require additional research.

Wilbanks suggested it is unlikely that a small set of societal indicators could be created that would be sufficient to the task. There is some thought that resiliency and vulnerability are too complicated. Health is an example. Health experts were asked if they could supply one health indicator that was influenced by climate change. The response was that no single indicator could incorporate the combined effects of exposures to allergens, pollens, the dangers of extreme weather, etc. It is not possible to establish an indicator simply for health, he said; it may be possible to create an indicator for something larger, but it would probably provide less information.

Wilbanks closed with three points: First, we can think about indicators based on observations of what society is doing for extreme weather events and water scarcity that is observed. What is happening now in changes in settlement patterns and land uses in areas that are almost certain to be more acute as climate change moves forward? We should not report speculations, but report the observation of changes that are occurring. Second, we should not rely on our "superior intellect" that makes perfect sense to us but turns out not to inform audiences at all. It is essential to have stakeholder participation and engagement. Third, he was engaged in working with insurance companies and financial institutions to establish that more resilient communities are lower risk communities and should therefore receive more favorable insurance and interest rates. The insurance and banking representatives contacted were not greatly impressed with the metrics presented. Therefore, the question was posed to them, "What data would you believe?" He regarded this as an important undertaking. These things are important as we try to help the NCA with societal indicators in moving toward reducing disruptions in the U.S. caused by climate change.

10.3 Carol Kramer-LeBlanc - U.S. Department of Agriculture (USDA) Sustainable Development

Carol Kramer-LeBlanc provided a summary of what the USDA is doing with respect to climate change. She summarized two conclusions of a 2008 assessment: first, that climate change is already affecting water resources, agriculture, and biodiversity; and second, that climate change would continue to exert such pressure. The USDA's vision is to help develop sustainable agriculture and forest systems that produce high quality food while reducing greenhouse gasses. She noted that the USDA has been engaged in climate research for more than 30 years and water research for more than 100 years, with a broad mission of technology transfer, management of public lands, and technical and educational assistance. The Department disseminates a broad range of statistics related to crops, forests, grasslands, soil types, management practices, and other matters. She emphasized that the USDA maintains agricultural land use data, which in many cases go back for decades and are available to provide baselines. It is difficult to get a handle on all the relevant indicators. Climate change is already affecting agriculture: crops are being produced in an atmosphere characterized by increasing carbon dioxide (CO_2) levels; livestock are also affected. She noted that agricultural systems are both sources and sinks of greenhouse gasses. One focus of the USDA climate change programs is to help produce agriculture and forestry systems that reduce CO_2. The USDA also maintains extensive relationships with land managers, researchers, state departments of agriculture, universities, the private sector, policy makers, and international organizations. She emphasized that weather and climate extremes are major limitations to production, land use change affects the environment, crops are being produced in an atmosphere with increasing CO_2, and livestock

and aquiculture are also affected by the effects of a changing climate.

Kramer-LeBlanc pointed out that a major international conference on the capacity of the planet to feed all its inhabitants had taken place in 2010 at The Hague. Given world population growth, she said, food security is an important social statistic; achieving the goal of sufficient food for the entire planet would be made more difficult by climate change. Food security and adequate nutrition are important indicators that reflect on the food system. She identified as important the efforts needed to change existing practices; for example, to reduce the use of petroleum-based fertilizers. There need to be improved ways to incentivize producers to adopt better practices to lower CO_2 and greenhouse gases. In conclusion, Kramer-LeBlanc called attention to current negotiations in global bioenergy partnerships as they relate to sustainability; these negotiations have created tentatively agreed upon social indicators. These related to land use; the price and supply of a nation's food basket; changes in income and employment in the bioenergy sector; changes in the unpaid time spent by women and children collecting biomass; changes in mortality due to cook stove smoke; the incidence of occupational death and injury, and others.

10.4 Jim Buizer - Science Policy Advisor to the President, Arizona State University

Jim Buizer reflected on discussions at the workshop. First, clarity is needed as to what people want from indicators, which means that the NCA needs to pay attention to its audience. Data need to be communicated in such a way that people notice and pay attention, trust them and believe there is something they can do with the data and information. It is not enough to analyze data; it is more important to induce action. Just because people have been warned does not mean they will act rationally. Buizer urged the group to start identifying what questions need to be answered and then work back to the indicator set and approach that would supply those answers. We should not feel compelled to pick only one indicator approach to address all questions. A mix of approaches can be used. He warned against assuming "trickle down" resiliency: those best equipped to get and use information tend to be those with the resources needed to create alternatives for themselves. It does not follow that this capacity will eventually work its way down to other people. There are inequities in capacities to implement alternatives. As a generalization, he urged those present to start with the indicators that are doable, scientifically defensible, and address issues that are international in scope. He pointed out that markets are global and that the NCA indicators effort is relevant to foreign policy responsibilities. He stressed that other countries have knowledge and experience that should be called upon. We should pay attention to what has already been done and what is known. Data can be misused, so stakeholders need to be engaged early and often and a two-way conversation should be maintained to ensure that the indicators are useful and meet the goals. The NCADAC should consider linking with federal agency efforts under the Interagency Climate Change Adaptation Taskforce.

Finally, Buizer urged that the indicator set for the 2013 NCA report should be representative, not comprehensive. Moreover, indicators should not focus solely on the negative factors associated with climate change, but on positive factors and opportunities as well. Furthermore, indicator development should start at the local level and scale upward. He recommended inclusion of an evaluation of the indicator system over time to determine if it is working, if it is being communicated and noticed, if it is being believed, and if it is resulting in action.

11 CONCLUDING REMARKS

The societal indicators workshop solicited inputs from a range of government and nongovernment experts regarding responses to the proposed NCA indicator goals, audience and scope, as well as input on the best practices and lessons learned to consider if the NCA develops indicators. To add specificity, the participants focused on the development of NCA climate-focused indicators for: (1) climate impacts and opportunities, (2) climate adaptation, (3) climate vulnerability and resilience, and (4) climate-related disaster preparedness. Moving forward, the workshop participants emphasized that it will be necessary to merge the lessons learned from the ecological, physical, and societal indicators workshops in developing a single, cohesive NCA indicator framework. During the workshop discussions, a number of points emerged as key messages worth considering as the NCA moves forward in developing an indicator framework:

- Indicators developed or selected for the NCA

should motivate the audience to notice and pay attention, believe the information, and do something about it.

- The NCA should start with the questions to be answered and then choose the indicators to best address the question.

- The NCA should draw lessons from and, where appropriate, build upon the many other indicators and indicator approaches that have been developed to address similar issues, as reviewed in the workshop. The indicator approach (e.g., composite, basket, and accounting) does not need to be the same for all of the indicator categories.

- The NCA should start with what is doable (i.e., "low hanging fruit"), especially in the short-term, and leverage existing efforts when possible.

- Indicators developed or selected for the NCA should be scientifically defensible, meet NCA peer-review standards, and be transparently presented in message, approach, and data sources.

- The NCA should engage stakeholders early and often in a two-way conversation, remembering that not all stakeholders are the same.

- The NCA indicator framework should be flexible, customizable, and serve multiple audiences in a way that builds common understanding among different groups.

- The process for selecting and developing indicators should include "citizen science" and experiential knowledge approaches.

- The indicators developed or selected for the NCA should be representative, not comprehensive (especially in the short-term).

- The indicators developed or selected for the NCA should reflect both negative and positive aspects of climate (i.e., impacts and opportunities, vulnerabilities and resiliencies).

- The indicators need to have appropriate coverage and be consistently gathered.

- The indicators selected should have enough frequency and consistency to be measured over time.

- The indicators developed or selected by the NCA should be evaluated and adaptivelymanaged to allow for changes over time.

Appendix A: Workshop Agenda

Wednesday, April 27, 2011

5:30   Meet and Greet Happy Hour

Thursday, April 28, 2011

Introduction to the Meeting

8:30   Welcome

        Lawrence Friedl, Acting Director of NASA Applied Sciences Program, NASA Earth Sciences Division

8:40   Introduction of Workshop Steering Committee and Workshop Logistics

        Workshop Co-chairs: Robert Chen (Columbia University), Melissa Kenney (AAAS Fellow/NOAA), Jim Smoot (NASA)

National Climate Assessment and Societal Indicators

8:45   The National Climate Assessment and Key Outcomes of the Ecological Indicators, Physical Indicators, Vulnerability and Valuation Workshops

        Kathy Jacobs, Director of National Climate Assessment, Office of Science and Technology Policy

Question and Answer Panel:

        Kathy Jacobs (Director of the National Climate Assessment)

        Emily Cloyd (NCA Ecological Indicators workshop)

        Fred Lipschultz (NCA Physical Indicators workshop)

        Julie Maldonado (NCA Vulnerability workshop)

        Fran Sussman (NCA Valuation workshop)

9:45   White Paper Summary and Charge to the Workshop Participants

        Melissa Kenney, AAAS Science and Technology Policy Fellow, NOAA Climate Program Office and Assistant Research Scientist, Johns Hopkins University

        Robert Chen, Director and Senior Research Scientist, Columbia University Center for International Earth Science Information Network (CIESIN) and Manager, NASA Socioeconomic Data and Applications Center (SEDAC)

        Sandra Baptista, Senior Staff Associate, Columbia University Center for International Earth Science Information Network (CIESIN)

10:30   Break

Approaches to Developing Societal Indicators

11:00   Panel Discussion on Types of Indicator Approaches Relevant to NCA with Q&A and discussion (10 minutes per panelist) Moderator: Caitlin Simpson

        Tom Wilbanks, Corporate Research Fellow, Oak Ridge National Laboratory

        Pat Gober, Director, Decision Center for a Desert City

        Mike McGeehin, Senior Epidemiologist, RTI International

Ben Campbell, Director of Environmental and Social Assessment, Millennium Challenge Corporation

Gemma Cranston, Lead Scientist, Global Footprint Network

Radley Horton, Associate Research Scientist, Columbia University/NASA GISS

12:15   Lunch (on your own)

Discussion: Indicator Framework and Approaches

1:30   Breakout Session Charge

Melissa Kenney, AAAS Science and Technology Policy Fellow, NOAA Climate Program Office and Assistant Research Scientist, Johns Hopkins University

1:45   Breakout Session #1 (4 groups, each addressing the same questions)

1. Goal for the NCA indicators.

2. Audience.

3. Scope.

3:00   Break

3:15   Breakout Session #2 (4 groups, each addressing the same questions)

1. Pros/Cons and Lessons Learned.

2. Must have Topical Societal Categories.

3. Input to NCADAC.

5:00   Adjourn

Friday, April 29, 2011

8:30   Report out from Breakout Sessions #1 and #2 from Thursday

Moving Forward with Societal Indicators: Categories, Requirements, Data, and Priorities

9:00   New Breakout Session Charge

Robert Chen, Director and Senior Research Scientist, Columbia University Center for International Earth Science Information Network (CIESIN) and Manager, NASA Socioeconomic Data and Applications Center (SEDAC)

9:15   Breakout Session #3 (4 groups, each addressing a different indicator question)

1. Climate impacts indicators.

2. Climate adaptation indicators.

3. Climate vulnerability and resilience indicators.

4. Climate preparedness indicators.

12:00   Lunch (on your own)

1:15   Report out from Breakout Session #3

The Path Forward – Priorities for the National Climate Assessment

1:45   Final Panel with questions and answers/discussion (10 minutes per panelist) Moderator:
       Bob O'Connor

       Dave Cleaves, Climate Change Advisor, USDA-Forest Service

       Tom Wilbanks, Corporate Research Fellow, Oak Ridge National Laboratory

       Carol Kramer-LeBlanc, Director, Sustainable Development

       Jim Buizer, Science Policy Advisor to the President, Arizona State University

2:45   Final Comments from the National Climate Assessment and Workshop Organizers

       Kathy Jacobs, Director of National Climate Assessment, OSTP

       Workshop Co-chairs: Robert Chen (Columbia University), Melissa Kenney (AAAS Fellow/NOAA),
       Jim Smoot (NASA)

3:00 Adjourn

Appendix B: Participant List

Sandra Baptista, Columbia University Center for International Earth Science Information Network

Nancy Beller-Simms, National Oceanic and Atmospheric Administration

Christa Blaisdell, U.S. Global Change Research Program

Robin Bronen, University of Alaska, Fairbanks

James Buizer, Arizona State University

Ben Campbell, Millennium Challenge Corporation

Robert S. Chen, Columbia University Center for International Earth Science Information Network

David Cleaves, U.S. Forest Service

Emily Cloyd, U.S. Global Change Research Program

Chelsea Combest-Friedman, National Oceanic and Atmospheric Administration

Gemma Cranston, Global Footprint Network

Mark Dunning, CDM Consulting Group

Anne Elixhauser, Agency for Healthcare Research and Quality

Barbara Entwisle, University of North Carolina at Chapel Hill

Mary Floyd, Zantech IT Services, Inc.

Lawrence Friedl, National Aeronautics and Space Administration

Patricia Gober, Decision Center for a Desert City

Bryce Golden-Chen, U.S. Global Change Research Program

William Goran, U.S. Army Corps of Engineers

Elisabeth Graffy, University of Wisconsin-Madison

David Hastings, National Oceanic and Atmospheric Administration

Radley Horton, Columbia University; National Aeronautics and Space Administration

Joe Hostler, Yurok Tribe Environmental Program

Nathan Hultman, University of Maryland

Kathy Jacobs, White House Office of Science and Technology Policy

Melissa Kenney, National Oceanic and Atmospheric Administration; Johns Hopkins University

Carol Kramer-LeBlanc, U.S. Department of Agriculture

Allison Leidner, National Aeronautics and Space Administration

Igor Linkov, U.S. Army Engineer Research and Development Center

Fred Lipschultz, U.S. Global Change Research Program

Susan Lovelace, National Oceanic and Atmospheric Administration

George Luber, Centers for Disease Control and Prevention

Julie Maldonado, U.S. Global Change Research Program; American University

Michael McGeehin, Research Triangle Institute

Susanne Moser, Susanne Moser Research & Consulting; Stanford University

Sheila O'Brien, U.S. Global Change Research Program

Robert O'Connor, National Science Foundation

Jennifer Parker, Centers for Disease Control and Prevention

John Pine, Appalachian State University

Dale Quattrochi, National Aeronautics and Space Administration Marshall Space Flight Center

Arthur Rypinski, U.S. Department of Transportation

Shubhayu Saha, Centers for Disease Control and Prevention

Caitlin Simpson, National Oceanic and Atmospheric Administration

Adam Smith, National Oceanic and Atmospheric Administration

John Smith, U.S. Geological Survey

James Smoot, National Aeronautics and Space Administration Marshall Space Flight Center

Fran Sussman, ICF International

Ariana Sutton-Grier, National Oceanic and Atmospheric Administration

Kimberly Thigpen Tart, National Institute of Environmental Health Sciences

Bob Vallario, U.S. Department of Energy

Anne Waple, National Oceanic and Atmospheric Administration

Linda Wennerberg, National Aeronautics and Space Administration

Thomas Wilbanks, Oak Ridge National Laboratory

Olga Wilhelmi, National Center for Atmospheric Research

Rob Winthrop, U.S. Bureau of Land Management

Steve Young, U.S. Environmental Protection Agency

Appendix C: Workshop Steering Committee Members

Co- Chairs

Robert S. Chen, Columbia University Center for International Earth Science Information Network, National Aeronautics and Space Administration Socioeconomic Data and Applications Center *

Melissa Kenney, AAAS Science and Technology Policy Fellow, National Oceanic and Atmospheric Administration, Johns Hopkins University*

Jim Smoot, National Aeronautics and Space Administration Marshall Space Flight Center Earth Science Office

Members

Sandra Baptista, Columbia University Center for International Earth Science Information Network*

C. Frank Casey, U.S. Geological Survey

Emily Cloyd, U.S. U.S. Global Change Research Program

Allison Leidner, AAAS Science and Technology Policy Fellow, National Aeronautics and Space Administration

Fred Lipschultz, U.S. Global Change Research Program

Julie Maldonado, U.S. Global Change Research Program, American University*

Gino Marinucci, Centers for Disease Control and Prevention

Bob O'Connor, National Science Foundation

Dale Quattrochi, National Aeronautics and Space Administration Marshall Space Flight Center Earth Science Office*

M. Carla Roncoli, University of Georgia

Shubhayu Saha, Centers for Disease Control and Prevention

Caitlin Simpson, National Oceanic and Atmospheric Administration Climate Program Office

Sue Stewart, U.S. Forest Service Research

Juli Trtanj, National Oceanic and Atmospheric Administration Oceans and Human Health Initiative

* Members of the Workshop Synthesis Team

Part 2: White Paper on the Development of Societal Indicators for the National Climate Assessment

Written by: Robert S. Chen, Melissa A. Kenney, and Thomas J. Wilbanks

With input from: Katharine L. Jacobs, Sandra R. Baptista, C. Frank Casey, Emily Cloyd, Robert E. O'Connor, Bryce Golden-Chen, Allison K. Leidner, Fred Lipschultz, Julie Maldonado, Gino Marinucci, Sheila O'Brien, Dale Quattrochi, M. Carla Roncoli, Shubhayu Saha, Caitlin Simpson, James L. Smoot, Susan I. Stewart, and Juli Trtanj

1 BACKGROUND AND PURPOSE OF INDICATORS FOR THE NATIONAL CLIMATE ASSESSMENT

1.1 National Climate Assessment

The National Climate Assessment (NCA) is being conducted under the auspices of the U.S. Global Change Research Program (USGCRP), pursuant to the Global Change Research Act of 1990, Section 106, which requires a report to Congress every 4 years. The NCA report

- "integrates, evaluates, and interprets the findings of the Program [the USGCRP] and discusses the scientific uncertainties associated with such findings;
- analyzes the effects of global change on the natural environment, agriculture, energy production and use, land and water resources, transportation, human health and welfare, human social systems, and biological diversity; and
- analyzes current trends in global change, both human-induced and natural, and projects major trends for the subsequent 25 to 100 years."

The current NCA (http://globalchange.gov/what-we-do/assessment/) differs in multiple ways from previous U.S. climate assessment efforts, being: (1) more focused on supporting the Nation's activities in adaptation and mitigation and on evaluating the current state of scientific knowledge relative to climate impacts and trends; (2) a long-term, consistent process for evaluation of climate risks and opportunities and providing information to support decision-making processes within regions and sectors; and (3) establishing a permanent assessment capacity both inside and outside of the federal government.

The NCA will therefore be an ongoing process that draws upon the work of stakeholders and scientists across the country. Assessment activities will result in the capacity to do ongoing assessments of vulnerability to climate stressors, observe and project impacts of climate change within regions and sectors, develop consistent indicators of progress in adaptation and mitigation activities, and allow for the production of a set of reports and Web-based products that are useful for decision making at multiple levels.

1.2 Purpose of indicators for the National Climate Assessment

The NCA vision for indicators is a small (less than 20), coordinated suite of climate-related physical, ecological, and societal indicators that both monitor key aspects of climate and climate impacts for the United States and are easily communicated to interested parties. These indicators will be tracked as a part of ongoing, long-term assessment activities, with adjustments as necessary to adapt to changing conditions and understanding.

The goals for the NCA indicators are to

- Provide meaningful, authoritative climate-relevant measures about the status, rates, and trends of key physical, ecological, and societal variables and values to inform decisions on management, research, and education at regional to national scales;
- Identify climate-related conditions and impacts to help develop effective mitigation and adaptation measures and reduce costs of management; and
- Document and communicate the climate-driven dynamic nature and condition of Earth's systems and societies, and provide a coordinated benchmark for all regions and sectors.

The NCA Indicators Workshops are part of a series of workshops intended to inform the process of developing indicators for the NCA to support monitoring, assessment, prediction, and decision making for the United States as it faces current and future effects of climate change. The participants at these workshops are charged with providing individual input to the Interagency National Climate Assessment (INCA) Task Force and the NCA Development and Advisory Committee (NCADAC; the federal advisory committee) that guide the NCA. The individual inputs from these three workshops have been or will be summarized in workshop reports to be provided to the NCADAC. Additionally, there will be a NCA indicator framework working group that will consist of a small group of participants from the ecological, physical, and societal indicators workshops that will consolidate the results of the three indicator reports into a white paper to provide NCA indicator framework options. The NCADAC will decide whether or not to pursue indicators for the NCA, and if they choose to do so, the indicator framework and its individual components will likely be developed by a NCADAC indicator working group.

Whereas other NCA workshops have focused on ecological and physical indicators, this workshop will 1) examine categories of societal indicators for the NCA, 2) explore alternative approaches to constructing indicators and their pros and cons for consideration for the NCA, 3) discuss specific requirements and criteria for implementing the indicators, and 4) suggest sources of data and potential contributors to such indicators. Societal indicators could include demographic, cultural, behavioral, institutional, economic, public health, and policy components relevant to impacts, vulnerabilities, and adaptation to climate variability and change as well as both proactive and reactive responses to climate variability and change. They should have clear links where appropriate to the physical and ecological indicators, but address not just what is happening to the environment, but also how human and societal systems are impacted by, preparing for, and responding to climate-induced environmental changes and to consider adaptation and mitigation strategies.

2 PURPOSE OF THE NATIONAL CLIMATE ASSESSMENT SOCIETAL INDICATORS

The intended foci of the NCA societal indicators are understanding, evaluating, communicating, and broadly informing decision making with regard to the status of the Nation in dealing with and preparing for climate variability and change. The following are the types of climate-focused questions that could be addressed with the indicators:

- How do we know that there is a changing climate? e.g.:
  o What are the climate's "vital signs" and how might they change?
  o How is the climate projected to change in the future?
- Are important climate impacts occurring or predicted to occur in the future? e.g.:
  o How can we (U.S. government, states, the public, etc.) tell if specific climate-related events, episodes, or trends are having significant economic, social, demographic, or other societal impacts, or not?
  o How can we tell if specific regions, sectors, or the Nation are being significantly affected by climate changes?
  o What are the most important impacts that are linked to climate change?
  o What is the anticipated rate of change?
- Are we adapting successfully? e.g.:
  o Is the U.S. adapting effectively to climate

variability and climate changes and associated impacts? If not, what are the consequences for other parts of the world?
  o Are other parts of the world adapting successfully? If not, what are the consequences for the U.S.?
  o Are adaptations keeping pace with impacts?
  o Are sufficient adaptation options available or under development to deal with anticipated future climate impacts given different levels of mitigation?
- What are the vulnerabilities and resiliencies given a changing climate? e.g.:
  o How can we tell if future vulnerability or resiliency to climate variability and change is increasing or decreasing in particular regions or sectors or the Nation due to climate adaptation and mitigation or due to non-climatic factors like migration, scientific and technological innovations, institutional changes, behavioral changes, and economic changes?
- Are we preparing adequately for future climate change? e.g.:
  o How can we tell what investments are being made to manage climate risks and if they are sufficiently effective and coordinated?
  o How can we tell if climate risks are increasing, decreasing, being shifted between regions, sectors, generations, or different elements of society?
  o What adaptation and mitigation scenarios and techniques need to be considered in response to climate change and variability?

It is not expected that the NCA societal indicators would be linked directly to a single decision or portfolio of decisions, but subsets of indicators, or the data supporting the indicator, might be used to inform decision-making processes such as the development and implementation of climate adaptation strategies in a particular sector or region.

2.1 What are the types of climate impacts of interest?
A wide range of climate impacts are of current and likely future concern, covering a range of sectors and topics, many of which have been addressed in more detail by other NCA workshops. Although it is beyond the scope of this white paper to review the range of potential impacts, it is useful to highlight some of the main areas of concern. Figure 1 presents an extract from a summary table from the

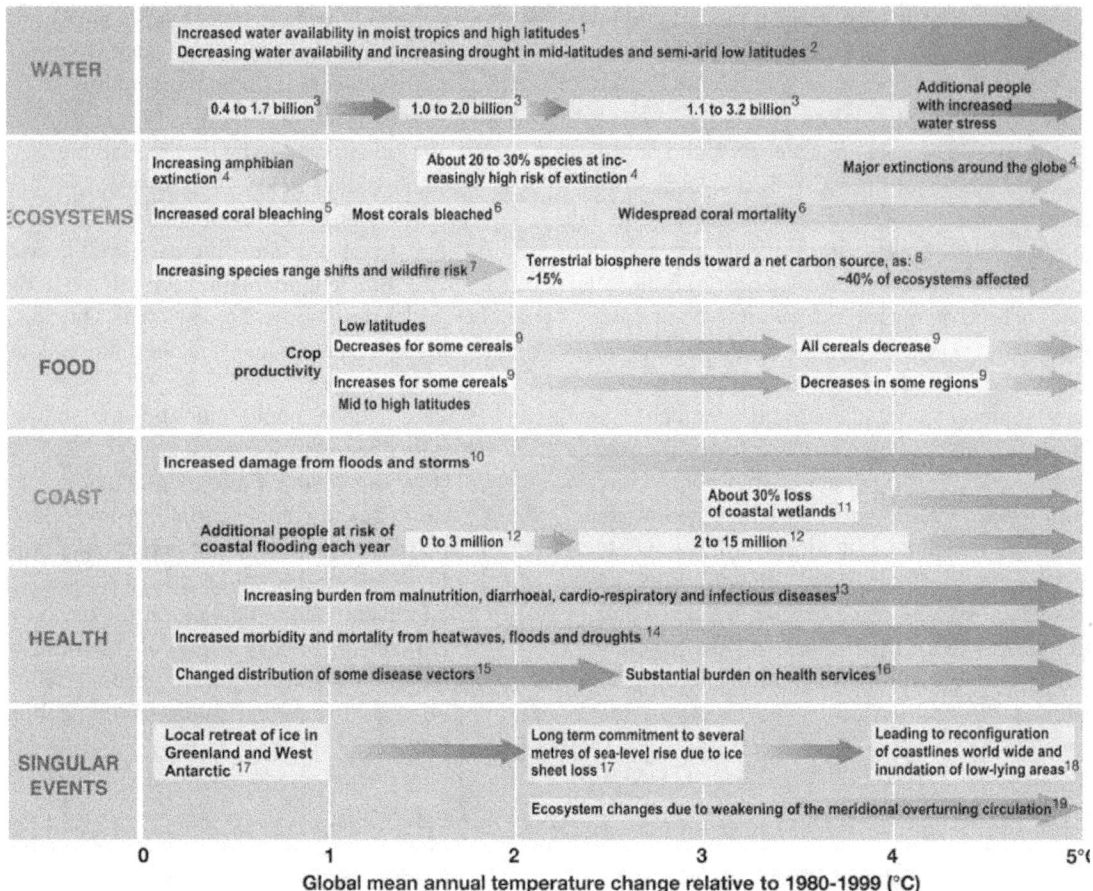

Table TS.3. Examples of global impacts projected for changes in climate (and sea level and atmospheric CO_2 where relevant) associated with different amounts of increase in global average surface temperature in the 21st century [T20.8]. This is a selection of some estimates currently available. All entries are from published studies in the chapters of the Assessment. (Continues below Table TS.4.)

Figure 1. Extract from Table TS.3 of the Working Group II report of the IPCC Fourth Assessment, Climate Change 2007: Impacts, Adaptation, and Vulnerability, illustrating example global climate impacts (IPCC, 2007).

IPCC Fourth Assessment Working Group 2 report on Impacts, Adaptation, and Vulnerability (IPCC, 2007), illustrating some of the major, better documented impacts associated with increasing levels of global mean surface warming.

Many of these types of climate impacts will result in multiple societal impacts, including changes in human mortality and morbidity, displacement of human populations, economic gains and losses, disruption of existing infrastructure, and changes in long-term environmental and economic sustainability of particular regions.

2.2 What are the types of adaptations of interest?

The National Research Council (NRC) report, *America's Climate Choices: Adapting to the Impacts of Climate Change* (2010), defines *adaptation* as the "adjustment in natural or human systems to a new or changing environment that exploits beneficial

opportunities or moderates negative effects" (NRC, 2010). Indicators of adaptation could help users understand the degree to which various regions or sectors are responding consciously or unconsciously to climate impacts, including both amelioration of adverse effects and exploitation of opportunities. The NRC report lists many possible adaptation options on the part of federal, state, local, private sector, nongovernmental, and individual stakeholders for major sectors including ecosystems, agriculture, forestry, water, health, transportation, energy, and oceans and coasts (NRC, 2010 Tables 3.2-3.8).

One major challenge is creating indicators of the effectiveness of alternative adaptation options, largely because we have not been able to observe the full results of the range of adaptation options implemented, and such adaptation measures are scale and impact dependent. Therefore, an alternative approach is to characterize actual levels of investment in adaptation. However, since adaptation investments may have different levels of success (and failure) over time, and may themselves have secondary impacts (e.g., changes in energy use due to improved building design or more reliance on air conditioning), measures of the effectiveness and secondary impacts of adaptation may also be important. For example, protective measures such as flood control in the Gulf Coast may have led over time to "maladaptive" agricultural and urban development in areas at high risk of storm surges due to tropical storms like Hurricane Katrina (NRC, 2006, 2010a).

2.3 How can current and future climate vulnerability and resilience be measured?
The 2010 NRC *America's Climate Choices* report defines vulnerability as "the degree to which a system is susceptible to, or unable to cope with, adverse effects of climate change, including climate variability and extremes. Vulnerability is a function of the character, magnitude, and rate of climate variation to which a system is exposed, its sensitivity, and its adaptive capacity." The white paper prepared for the NCA's Vulnerability Assessment workshop (Mills and Ebi, 2011) lists additional definitions of vulnerability and gives an overview of alternative vulnerability assessment approaches. For the purposes of the NCA, simple measures of climate vulnerability are preferable. For a relatively straightforward impact like sea level rise, it may be possible to quantify future vulnerability and resilience in terms of the elevation and slope of coastal regions, their existing and projected population and settlements, the current and future state of soil structure, water tables, and land cover, and current and planned protective infrastructure and ecosystems. Even in this case, there are likely to be numerous uncertainties related to the risks posed by extreme events, the rapidity of sea level rise, demographic and social changes, and the economic sustainability of settlements and built infrastructure.

From the viewpoint of society's heterogeneous ability to deal with change, regardless of the specific stressor, it may also be worth exploring more generic indicators of adaptive capacity (or its lack) such as measures of poverty, infant mortality, age structure, conflict, and government effectiveness.

2.4 How can we assess preparedness?
From a policy perspective, indicators of adaptation and vulnerability are only part of what most policy and decision makers need. They also need integrated information on how prepared a particular region, sector, or jurisdiction is to address climate variability and changes and associated impacts, and how preparedness and response capacities can be improved. The need for response capacity depends on whether adaptation is or will occur successfully, and at what cost, and on the types and levels of vulnerability that need to be addressed. A low-income region may be highly prepared for climate change if its natural ecosystems are robust and healthy, its institutions and governance practices are effective and just, its population is well-organized and has diverse income sources, and appropriate sources of data and information are readily available. A densely populated urban region may be highly prepared for a certain range of extreme events, but poorly prepared for a mega-disaster or an extended period of disruption. Relevant indicators might include emergency response capacity (per capita availability of hospital beds, ambulances, medical personnel, etc.), infrastructure flexibility (e.g., bottlenecks vs. alternate routes and transportation modes, access to external networks and supplementary sources of energy, water, food, and other supplies), financial stability and reserves (e.g., investor ratings), monitoring and assessment capacity with regard to climate change and impacts, and the knowledge and expertise of both personnel and the general public. Many of these factors are subject to direct policy intervention, and therefore indicators of preparedness are likely to be of high interest in policy and decision-making.

3 THE KNOWLEDGE BASE FOR CONSIDERING SOCIETAL INDICATORS

The search for societal indicators has been an active field of research and practice for more than four decades. In the United States, a strong interest in finding ways to measure and compare levels of human well-being dates back to the ambitious social programs of the 1960s. "Social Indicators" were conceived and developed for such concerns as population, health, education, security, living conditions, economic conditions/poverty, and political contexts. The journal *Social Indicators Research* was founded in 1974, and by the 1980s the collection of social indicator data had become common in national and international data systems, e.g., United Nations, Handbook of Social Indicators, 1989.

More recently, rather than working forward from available social data sources, there has been a growing interest in developing indicators of social or social-environmental contexts, e.g., indicators of *vulnerability* to climate change impacts or other environmental threats (e.g., Moss et al., 2001; Eriksen and Kelly, 2007), indicators of *sustainability* of human and/or natural systems or the lack thereof (e.g., NRC, 1999; NRC, 2010b), and indicators of *resilience* as a way to reward high levels and seek to address low levels (e.g., Cutter et al., 2010; Sherrieb, Norris, and Galea, 2010). Although the state of this science/art is advancing gradually, the general view is that currently available indicators are not yet very satisfactory, for instance as a basis for making decisions about the allocation of resources in order to improve capacities that appear problematic. In fact, discussions of research needs in these fields nearly always identify indicators as a high priority for research.

Running through this body of knowledge and experience are several challenges. First, in many cases important dimensions of social systems lack time-series data to support robust examinations of trends. Either data about such dimensions have not been gathered in the past, or data are exceedingly crude, e.g., one observation every ten years from the decadal census. Second, what is relatively easy to measure is not necessarily a true indicator of what one would like to know. For instance, measuring the resilience of a community depends on gauging social dynamics rather than socioeconomic characteristics of well-being: how does one measure "connectedness" or capacities for community problem solving? As another example, the proportion of a population that has achieved each of a set of educational levels is usually feasible to measure, but does that really indicate what people know? Or what their capacities are for adaptive problem solving? Third, where the questions pertain to nature-society relationships, social data need to be integrated with natural science data, raising issues such as differences in units of measure, scales of data aggregation, and simply a lack of bridge-building expertise. Examples of recent U.S. government experience with such challenges include Admiral Lautenbacher's effort to connect earth-observing systems from space with "social benefits" objectives (GEOSS; see also the NASA "decadal study": NASA, 2007) and discussions by the National Academy of Sciences (NAS)/NRC Committee on Human Dimensions of Global Change of strategies for earth-observing systems in *situ* (such as the National Ecological Observatory Network (NEON) and the U.S. Long Term Ecological Research (LTER) Network) to link environmental observations with socioeconomic indicators. The way has been neither easy nor smooth.

This rich and diverse combination of societal indicators research and practice suggests several insights for thinking about societal indicators in the NCA:

a. To get the right indicators, it is essential to work back from the important questions to be answered, rather than starting with readily available data sources. Even if certain dimensions are very difficult to observe and measure in practice, the development of estimates or proxies should be linked as closely as possible to the need.

b. Because it is so easy to lose one's way in a mass of possible measures, it is important to focus on high-priority societal issues, which in the case of NCA means salient societal consequences of first-order climate change effects, e.g.:

(1) Stresses of temperature, precipitation, severe weather, and sea-level changes – in both averages and extremes – for societal systems (in a multi-stress context), related to serious challenges to human well-being and social stability in especially vulnerable situations

(2) Early warnings about emerging problems to inform timely policy responses, especially where tipping points/threshold effects might be a factor

c. There is no one set of indicators that are equally good for all purposes: contexts matter. Because threats differ, locations differ, scales differ, and sectors differ, it is often desirable to think in terms of menus of indicators rather than a single small set.

d. It is highly useful to consult stakeholders in the early stages of designing indicator systems and mechanisms for packaging and supplying data, in order to increase the likelihood that indicators will be useful – and used. In this connection, the interests of NCA intersect with the interest in national climate services.

A final insight would be that in many cases the existing knowledge base does not support the development of valid indicators of what we want to know about (e.g., resilience). Arriving at the right set of societal indicators for the long-term NCA infrastructure is likely to require some gap-filling research and may require some new societal data systems.

4 INDICATOR DEFINITIONS AND SELECTION CRITERIA

4.1 Definitions
The NCA may use the term indicator and supporting terms differently than other groups. As a result, we use the following terms and definitions in this document.

- An **indicator** is a direct measure, proxy, or index that is used to understand, evaluate, and communicate the impacts and vulnerabilities resulting from climate change and variability. It is used to broadly inform decisions, but the NCA indicators are not intended to support a specific decision or a portfolio of decisions.
- An **index** is a constructed measure where multiple measured variables are combined to provide an assessment of an area of interest that cannot be adequately captured using a single measure or proxy (Keeney and Gregory, 2005).
- A **metric** or **measure** is a variable that is used individually or in combination with other data to quantify the indicator.

4.2 Qualities of a good indicator
Though there are some general qualities of a good indicator, there are also some qualities that may be especially important for the planned NCA set of indicators. For the NCA indicators, they cannot comprehensively address all potential questions; however, one of the most desirable qualities of the NCA indicators is that they be representative. Representative indicators address the most important climate-related impacts, vulnerabilities, adaptations, and preparedness. A representative set of indicators does not have to include all potential indicators, nor map perfectly to all individual sectors or regions, but it should provide an appropriate overview of current societal, ecological, and physical climate impacts and vulnerabilities/resiliencies as well as the effectiveness of current adaptation and preparedness efforts. In this regard, it is essential for the indicators to have an unambiguous, defensible linkage to climate variability and change.

Given a comprehensive set of indicators, it is essential to ensure that the component indicators are **analytically sound** (Schepelmann et al., 2010). Analytically sound indicators are those that are

based on a scientifically defensible theoretical framework and are transparent in their presentation of methods and data. The indicators must additionally have components that are **measurable** (Schepelmann et al., 2010). Indicators need to be based on data that are available (to both the NCA and the public), well-documented and peer-reviewed, and appropriate to include individually or in aggregate for a given indicator. The measurability criterion may also highlight key data gaps or existing data or indicator efforts that could be effectively leveraged by the NCA indicator system.

Additionally, it is important to have indicators that are **understandable**, meaning they are easily communicated and understood by a range of users with different levels of technical sophistication (Keeney and Gregory, 2005). It may be useful to involve a range of stakeholders in testing whether or not the indicators achieve this criterion. Similarly, **operational** indicators (Keeney and Gregory, 2005) are those that transparently describe and distinguish the scientific data and methods and the value judgments in the weightings such that a sophisticated user could understand the component parts and apply their own weightings, as appropriate.

Finally, in addition to understanding, evaluating, and communicating, it is essential that the set of NCA indicators be **policy relevant** (Schepelmann et al., 2010). The indicators could track the current state of adaptation or preparedness, or be used to assess changes in impacts and vulnerability/resilience given different mitigation and adaptation options, potentially through the use of scenarios. Some might focus on specific societal, ecological, or physical topic areas, whereas others could be designed to bridge across topic areas to characterize the impacts on the interconnected human-natural-physical system. Additionally, to the extent possible, it would be useful to understand the sensitivity of the indicators to various types of climate variability and change.

5 APPROACHES TO DEVELOPING INDICATORS

Many different indicators have been developed and used with varying degrees of success by both governmental and nongovernmental bodies to support understanding, communication, evaluation, and decision-making for a diverse set of societal issues. It is therefore useful in planning the NCA indicators to review key lessons learned from related efforts and to carefully match the goals and intended audiences for the NCA indicators with appropriate strategies and implementation approaches.

With this in mind, a detailed inventory of approximately 40 indicators is being developed (see Part 4: Societal Indicators Inventory), highlighting a diverse set of indicator approaches, temporal and spatial scales, and sectors and topics. Although not an exhaustive compilation, the inventory provides summary information on a range of indicators that have been used in a number of policy contexts, including references to key background and evaluative literature.

Based on this inventory and a review of the indicator literature, we suggest three main groups of indicator approaches that could be considered individually or in combination to develop indicators for the NCA:

     a. composite indicators and indices,
     b. dashboards and baskets of indicators, and
     c. "systems" or accounting indicators.

The following subsections summarize some of the main features of these alternative approaches and discuss a few examples from each category.

5.1 Composite indicators and indices

Composite indicators and indices encompass efforts to characterize the behavior of complex systems in a single quantitative measure (or very small set of measures) to enable simplified comparisons, tracking, and messaging. Many composite indicators are widely used and cited, including economic indicators like the Consumer Price Index (CPI) and the Dow Jones Industrial Average; development indicators like the Human Development Index (HDI); political indicators like the Corruption Perceptions Index and the Freedom in the World survey; and environmental indicators like the Environmental Vulnerability Index (EVI) and the Environmental Performance Index (EPI). Some composite indicators are based on complex data structures designed to characterize a wide range of system behaviors; others focus on foundational system elements viewed as drivers of system change or on "bellweather" elements that are sensitive to short- or long-term fluctuations (e.g., the prices of certain commodities or corporate stocks); and at least one prominent indicator, the Doomsday Clock of the *Bulletin of the Atomic Scientists*, is based entirely on the analysis and judgment of a distinguished group of experts.

An important premise of some composite indicators like the HDI is that there are some measurable phenomena in society that reflect the overall function—or dysfunction—of key systems. For example, the HDI is based on only three indicators: *(1)* life expectancy at birth, reflecting the projected survival rates of newborns based on current patterns of mortality; *(2)* an education index based on estimates of years of schooling; and *(3)* gross national income per capita.[1] Higher life expectancies are generally associated with better nutrition, health care, sanitation, security, etc.; longer education with greater access to and use of knowledge; and higher per capita income with greater financial security and resources. Overall development is thus reflected as progress across multiple, complementary dimensions.

Other indicator approaches such as the EPI and EVI tap very large sets of input data and indicators in order to assess a broad range of environmental and human system characteristics. Regardless of the number of input variables, methods for combining disparate types of data vary greatly and entail a number of decisions. One decision is to determine the relative weights assigned to different components, which is largely a normative choice reflecting the priorities or preferences of the index developer (or potentially the users, if the components can be unpacked to allow a sophisticated user to assign their own weights to the components). Such choices can strongly influence the resulting composite index and any derived rankings or categorizations.

5.2 Dashboards and baskets of indicators

Instead of combining disparate variables into a single composite indicator, some efforts have focused on developing larger baskets or dashboards of indicators that recognize the multi-dimensionality of systems and problems and the difficulties associated with combining often disparate component indicators. For example, some indicators may better reflect short-term, rapidly varying elements of a system, whereas others may capture long-term trends or shifts in spatial or temporal patterns. Separate indicators for different sectors or regions can allow for more detailed monitoring of interactions and feedbacks between sectors and regions, which would not be possible when indicators are aggregated. Moreover, the relative importance of different indicators may vary depending on the state of the system and the needs of users. Therefore, there may

[1] Prior to 2010, the HDI was based on Gross Domestic Product per capita.

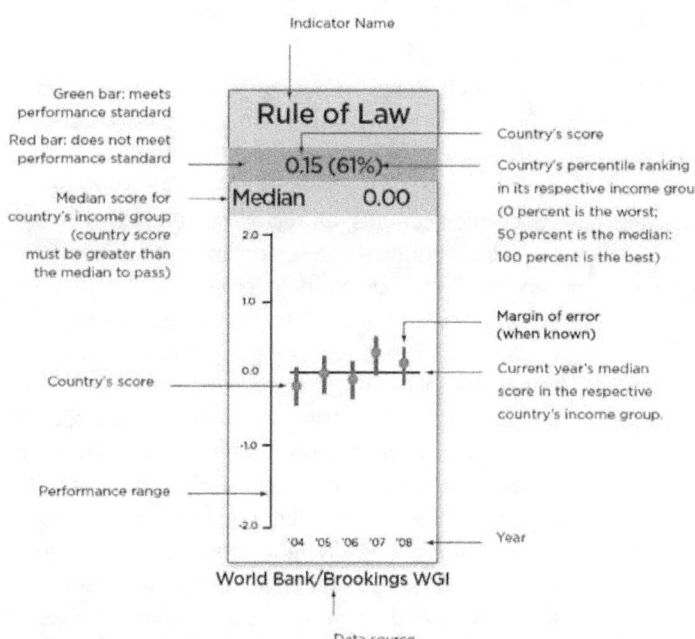

Figure 2. Example of a scorecard for a single indicator in the Millennium Challenge Corporation (MCC) basket of indicators (MCC, 2011). <http://www.mcc.gov/documents/reports/reference-2010001040503-_fy11guidetotheindicators.pdf>

not be a single set of weights appropriate to all users and applications.

One example of this approach is the Millennium Challenge Corporation (MCC) Indicators, a basket of 17 indicators developed by third parties covering three broad themes: Ruling Justly, Investing in People, and Encouraging Economic Freedom. The MCC uses these indicators to generate an annual "scorecard" which provides both comparisons with other countries as well as time trends (Figure 2). The MCC Board evaluates country performance based on these scorecards as part of the process of determining a country's eligibility for development assistance from the MCC.

Other examples of baskets of indicators include the European Sustainable Development Indicators, the World Water Development Report Indicators, and the just release "Pay Now, Pay Later" state-by-state reports (http://www.secureamericanfuture.org/pay-now-pay-later/). In the case of the World Water Development Reports, the basket of indicators varied considerably between the first, second, and third reports due to significant changes and inconsistencies in data availability, limiting the utility of the indicators for assessing changes over time. A United Nations Task Force is currently attempting to address this limitation.

Many of these efforts have explored creative ways to present multi-dimensional indicators in understandable forms, e.g., through "dashboards" that reflect the different types of data and indicators that one might need to drive a car or fly an airplane. It is clearly still a difficult challenge to provide clear, simple, and compelling messages based on these types of results to non-scientific audiences. On the other hand, specific indicators or subsets of indicators may be of particular interest for use within formal decision-support frameworks, as is the case with the MCC Indicators.

5.3 "Systems" or accounting indicators

A systems or accounting approach to indicators is based on identifying a common framework such as the national accounts system and corresponding unit of analysis or "currency" that can be used to translate impacts or activities across multiple sectors, regions, time periods, and other boundaries. For example, indicators such as the "Green GDP" and the Genuine Progress Indicator are based on adjusting national accounts data to reflect monetized costs associated with pollution, resource depletion, loss of ecosystems and biodiversity, and other environmental impacts. The Genuine Savings indicator developed by the World Bank estimates the effects of resource extraction and greenhouse gas emissions on the net savings rate of a country. Alternatively, the widely-used Ecological Footprint utilizes the area of biologically productive land and ocean per person as the basis for comparing the human use of resources with the planet's available carrying capacity. In 2007, the U.S. ecological footprint was estimated at 8.00 global hectares per person (gha), slightly more than double its biocapacity of 3.87 gha.

6 KEY ISSUES TO CONSIDER IN SELECTING AN INDICATOR APPROACH

6.1 Audience
Perhaps the most critical issue in selecting the NCA societal indicator approach is the intended audience: who are the indicators aimed at, and what is it they are expected to learn or gain from the indicators?

A primary audience for the NCA indicators is certainly the collection of present and future legislative and executive branch leaders at federal and state levels. A second major audience is the general public in the U.S., specifically the interested and informed public, to support the NCA communication strategy. Other organizations and individuals may find the indicators useful, but the primary intent is to develop indicators that communicate the status of the Nation with regard to progress in dealing with and preparing for climate change. The NCA indicators should help these audiences understand and evaluate complex and often confusing information about climate change impacts, by integrating diverse data and information into a finite set of quantitative measures using objective and transparent methods.

Put another way, the NCA indicators need to capture key messages about climate impacts, adaptation, and responses in simple terms understandable to the target audiences. In other indicator approaches, this is often done through analogy, e.g., threat levels and dashboards, or by comparative rankings. Rankings offer a way to call attention to differences in status or performance on the part of different regions or administrative units, but they are less useful in communicating absolute levels of impact or risk.

6.2 Scope

An ideal set of societal indicators would characterize the full range of climate-related impacts and vulnerabilities, societal adaptation, and efforts to mitigate climate change and current and future impacts. In practice, the diversity of potential climate impacts and responses and of societal outcomes of interest suggests that it will be difficult to find or construct a small set of indicators that will cover this range completely. Two alternative strategies are therefore to

- Identify or develop representative indicators that capture key processes or thresholds in society's response to climate change (e.g., the way the Infant Mortality Rate is used as a key indicator of human well-being and social system effectiveness); or
- Identify or develop a few systematic metrics that can be developed in consistent ways across regions and sectors to characterize the most significant outcomes of interest, e.g., costs and mortality.

6.2.1 INTEGRATION ACROSS SECTORS AND TOPICS

A major challenge in selecting societal indicators is how to decide what economic sectors or key topics should be covered directly, or indirectly, by the indicators. In a comprehensive system of economic accounts, all sectors would in theory be included, though some sectors might therefore have very small impacts on the overall indicator values. Selecting a few key sectors such as agriculture, energy, water, public health, and transportation has the benefit of highlighting key areas of impact, but may miss climate-sensitive activities in such sectors as forestry, fisheries, tourism, and housing. In some cases, climate changes manifested as changes in the frequency and/or magnitude of extreme events such as hurricanes, droughts, floods, and wildfires will have cross-sectoral impacts that are closely intertwined and result in net effects that are different from the individual sectoral impacts.

6.2.2 SPATIAL SCALES

The primary objective of the NCA societal indicators is to provide information useful for the Nation as a whole, but an important issue is how different regions of the U.S. might be affected in different ways, to different degrees, and at different times. Specifically, averaging across spatial scales could completely eliminate important information about important climate impacts present at smaller spatial scales; so finding the balance between aggregation and regional specificity is critical. An ideal approach is therefore to develop "nested" indicators so that the same indicator approach can be used at local, regional, and national scales to allow for comparisons across geography and targeted identification of vulnerable locations. One complication is that for some types of climate impacts such as water stress or air quality, indicators based on standard administrative units such as states may be less informative or useful than indicators based on alternative regional units (e.g., watersheds, airsheds, or mega-regions). Considering regions such as the Southeast and Gulf Coast, Alaska, Hawaii, the 'Bos-Wash' urban corridor (Florida et al., 2008), and areas bordering Canada and Mexico, some indicators may have particular relevance for cross-boundary topics including trade routes and patterns; water, food, and energy security; and international and transboundary agreements.

A second complication relates to the global nature of the economy as well as international flows of people, pests, disease, and information. Climate impacts and responses in other parts of the world could have significant secondary effects on the U.S. or specific sectors or regions, including impacts on trade, tourism, migration, prices, and public health. Characterizing these interrelationships is essential, as otherwise key adaptive responses such as coordinated disaster preparedness and response, trade, outsourcing, and migration may be missed.

6.2.3 DIAGNOSIS VS. PROGNOSIS

A primary goal of the societal indicators is to assess whether or not climate impacts and adaptations are currently occurring, and how significant these impacts are relative to the past. However, to support decision-making in a changing climate, it may also be important to assess future trends in impacts, vulnerability, and adaptive capacity. On short time scales, this could take the form of "leading" indicators; that is, indicators that are based on parts of the system or regions of the Nation (or other parts of the world) that may be more sensitive to certain climate factors (e.g., the tourism sector or permafrost regions in Alaska). On longer time scales, there may be impacts and vulnerabilities arising from climate changes already expected to occur but not yet fully realized due to lags in climatic and environmental systems (e.g., sea level rise or changes in ecological zones). Indicators could be projected given different climate scenarios to help decision makers think about the impact of climate given different future conditions.

For example, a plausible impact indicator might be the total population currently living in areas likely to face inundation within, say, fifty years due to projected sea level rise based on environmental changes already under way as the result of greenhouse gas emissions. This indicator would vary over time based on progress or lack of progress in controlling emissions, the predicted response of the oceans and cryosphere to warming during the next fifty years, and population change in the affected area. It could help planners and policy makers assess current levels of vulnerability to sea level rise and guide decisions about protective infrastructure, insurance approaches, land use planning, and other responses.

6.2.4 TIME SCALES (AVERAGING, REPEAT, AND LEAD TIMES)

In addition to the issue of diagnosis versus prognosis, there are also other important issues related to the time scales of the indicators. From the viewpoint of policy-making and communication, it may be important to issue indicators on a regular basis, e.g., as often as annually. Many but not all social, economic, and environmental datasets are available on an annual basis. However, in some cases, there is considerable annual variability (e.g., in the costs and mortality due to extreme weather events), so that providing averaged or smoothed data to emphasize long-term trends in impacts is warranted.

As in the case of physical climate indicators, tracking long-term changes in the *variability* of impacts may also be very interesting as an indicator of possible changes in the sensitivity or adaptive capacity of societal systems with respect to changing climate or changing climate variability. For example, increases in the number of extremely hot days might lead to some adaptations (such as more use of air conditioning) that could increase peak energy demand and other adaptations (such as better insulated buildings) that could reduce variability in energy demand. Both base and peak energy demand are key factors in thinking about future vulnerability and adaptive capacity in the energy sector.

There are also likely to be major differences in data availability and its regional scale for different types of indicators, e.g., lags of months to years due to data collection, processing, and analysis activities, such as for the U.S. Decennial Census. Another important concern is valuation of potential future impacts versus near-term impacts—for example, how can the impacts of future loss of coastal land to sea level rise decades from now be compared quantitatively with medium-term losses due to changes in coastal storm frequency and magnitude?

From this perspective, a basket of indicators approach may have the advantage of allowing for some diversity in the time scales of indicators, better reflecting the different time frames and dynamics of climate impacts, adaptations, and response. Formulation of a composite index or a consistent systems framework requires selection of a specific approach to translate indicators or variables with diverse time scales into equivalent "present-day" terms. For example, this could entail use of a "discount rate" applied to monetary valuations, a potentially contentious issue with normative implications. Though the indicators can be updated at different times given the indicator and its data sources and decision-relevant timeframe for that indicator, it is essential to update all the indicators at some regular time frame, such as every four years for the NCA report, to provide an update that can have a large communication impact and provide a comprehensive synthesis.

6.2.5 DETECTION AND LINK TO CLIMATE CHANGE AND VARIABILITY

A key issue in designing and selecting societal indicators is the degree to which significant change—or lack of change—can be reliably detected and, where appropriate, linked to a climate-related cause or driver. One necessary ingredient for detection is the availability of sufficient historical data to establish a suitable baseline so that short-term variations can be distinguished from long-term trends. For many indicators, adequate spatial coverage or sampling is also needed to ensure that changing spatial patterns are not mistaken for aggregate trends and also to detect significant changes in the spatial distribution of impacts and vulnerability. Strong linkages to climate variability and change may require significant amounts of related data on comparable spatial and temporal scales in order to sort out confounding factors and to establish clear associations or causal relationships.

As an example, an agricultural yield indicator based on all food crops might be a good overall indicator of changes in national agricultural productivity, but spatial and temporal variations would be expected both because of climate trends and extremes and as the result of changes in management, technology, economics, and policy (e.g., incentives for biofuels). An indicator based only on climate-sensitive crops might make establishing the linkage to climate easier, but could miss important impacts in crops not normally considered sensitive to climate, impacts resulting from indirect effects (e.g., markets for food or biofuels), or impacts resulting from changes in the mix of crops or the introduction of new crops. Another complication for establishing strong linkages to a changing climate relates to potential non-linear effects associated with multiple direct and indirect climate changes (e.g., changes in plant evapotranspiration associated with increased atmospheric CO_2 concentrations, relative humidity, temperature, solar insolation, and windiness).

6.3 Transparency and validation

The credibility of the NCA societal indicators will depend substantially on the transparency of the process of developing and maintaining the indicators and the degree to which the indicators can be validated. The rationale for selecting specific indicators and input variables needs to be clearly articulated, along with a strategy for assessing the significance of observed changes relative to known sources of error and uncertainty. Providing the indicator methods and data in the technical supporting information will be key to ensuring transparency of the indicator method, assumptions, data sources, and uncertainties. The NCA can provide this information and data via the Web interface; such presentation methods may drive priorities for the ongoing process.

6.3.1 TREATMENT OF UNCERTAINTY

All observations of physical and social phenomena inherently involve uncertainties, and uncertainties may be increased or reduced through sampling, aggregation, transformation, analysis, modeling, and interpretation. In many instances, tracking changes over time entails fewer uncertainties than estimating absolute levels of a parameter: for example, many economic indicators are presented as percentage changes from a reference year or period. Ordinal indices or ranks permit more general comparisons of relative status or activity of different groups or regions, taking into account uncertainties in component indicators and input data. In many practical warning systems, quantitative indicators are translated into a small number of categories (e.g., high-medium-low or red-yellow-green), which effectively decreases the impact of uncertainties within the categories but may increase their impact near the category boundaries or thresholds.

In this regard, careful attention needs to be given to the tradeoffs between the likelihoods of "false positives" versus "false negatives" in defining categories and thresholds, taking into account the purpose of the indicators. For example, to ensure that the target users are adequately forewarned when climate impacts are becoming significant, warming thresholds may need to be set "low" (e.g., at a low level of statistical significance or when only limited data are available), but this may increase the number of warnings that turn out not to be significant, perhaps resulting in "warning fatigue" or perceptions of a "cry wolf" syndrome. Alternatively, to avoid excessive false positives, warning thresholds could be set "high" (i.e., a high level of statistical significance and more complete data), but this increases the risk that warnings or diagnoses may be provided too late to be useful to target users.

6.3.2 RELIABILITY/REPUTATION OF DATA SOURCES AND CONTROLS/CHECKS FOR POSSIBLE BIAS

Clearly, an important issue for any indicator system is the reliability and quality of the data sources and the process by which errors, biases, and other problems are identified and addressed. Many sources of socioeconomic data collect data for administrative or regulatory rather than scientific purposes, so that careful attention is needed to address problems such as incomplete reporting, different response rates across different groups, incentives to under- or over-report, and fraudulent submissions. For example, disaster loss estimates are often biased on the one hand by the desire to inflate damage estimates to qualify for disaster assistance or insurance payments and on the other by varying definitions of and ability to measure direct and indirect losses.

6.3.3 POTENTIAL INVOLVEMENT OF STAKEHOLDERS AND USERS

The indicator selection process may involve key stakeholders and user groups (Morin, 2005); indeed, the NCA strategy anticipates such engagement. The involvement of stakeholders in developing the indicator framework can help support the use of indicators as a communication tool and increase the use and usefulness of the indicators. The inclusion of stakeholders in the initial indicator development and selection process, when their input is able to influence the design of the framework, can increase stakeholder buy-in, build capacity in the expert and information networks, and help to meet the NCA public engagement goals.

A priority for the NCA indicator system is to communicate climate impacts and vulnerabilities to the public and decision makers. Thus, it is critical for stakeholders and users to review and comment on the indicators for their clarity as a communication tool and their effectiveness in broadly informing understanding of the impacts related to important climate-sensitive decisions. Additionally, multiple approaches (i.e., both aggregated and disaggregated indicators in the report and on the NCA's Web-based component) and presentation methods (e.g., single value, map, score card, and graphics) can be provided to stakeholders and users to test which approach best matches user needs and could be adapted to inform adaptation decisions.

Stakeholders can also articulate the potential uses and usefulness of the indicators inside and outside of the NCA. Other groups or individuals could adopt the indicators for the NCA to inform adaptation decisions, understand the climate-sensitivity of multi-stressor impacts, or to explore the impact of policy options to improve certain indicators. After the indicator framework is implemented, continuing engagement of scientists and stakeholders will help the NCA periodically assess the uses of the NCA indicators and the gaps in near-term or future efforts to improve the usefulness of the indicators.

6.3.4 INCORPORATION OF PREFERENCES AND NORMS

Any selection of indicators will implicitly involve some degree of judgment about the relative importance of different factors for the Nation, especially current impacts and costs versus possible or expected future impacts and costs. How much weight should we give to impacts born by future generations—which may or may not come about and for which they may or may not be better prepared than us to deal with—compared with impacts experienced now or in the near future? This is basically a normative decision, which social scientists have tried to quantify using economic or social discount rates of various types.

Similarly, in developing composite indices, either explicit or implicit decisions are needed about the relative weights to be assigned to different component indicators or dimensions. Often, these are based on normative judgments about, for example, the relative importance of local versus regional versus global problems or short-term versus long-term effects. In most cases, we lack sufficient data on past "good" or "bad" outcomes to test or validate the selection of weights against historical experience.

The basket of indicators approach arguably gives users the greatest flexibility to apply their own normative preferences to the selection and use of indicators, within the limits of the set of indicators and component data provided. In the case of composite indicators, it is not difficult to make the component data and weights available to users and indeed to provide interactive analysis and visualization tools to allow users to choose their own set of weights or combine the components in different ways. This would provide more transparency with regard to the incorporation of norms, and give users the opportunity to customize the indicators to take into account their own preferences and discount rates and address their specific questions or decision-support needs.

6.4 Usability

Usability encompasses a range of desired characteristics of indicators that affect how the target audience is able to understand, interpret, and act upon the information embodied in the indicators.

6.4.1 ABILITY TO INCORPORATE INDICATORS INTO SCENARIOS AND MODEL PROJECTIONS

If we are interested in developing prognostic indicators that will project changes in the indicators that would result from different futures, then these indicators will need to have clear ties to available prediction models for climate and climate impacts. For example, it may be possible to develop a broad set of indicators of mortality and morbidity due to climate variability and change based on currently available data, but more difficult to project the full range of public health impacts in the future. It may therefore be desirable to select a smaller set of representative indicators for which model predictions are feasible, e.g., mortality due to extreme heat or cold episodes. It would be useful to begin conversations early in the development of the indicator framework to explore opportunities and limitations in linking indicators with climate impact models. Consistency between observed and projected indicators will also help users bridge the gap between a diagnostic understanding of the current state of climate impacts, adaptation, vulnerability, and preparedness and a prognostic view of possible future changes in these dimensions based on current understanding.

6.4.2 ABILITY TO LINK INDICATORS TO KEY POLICY LEVERS

When designing indicators, one of the opportunities is to explore the inclusion of indicators that characterize the degree to which different policy options are being implemented and their performance to date. Since a goal of the NCA indicators is to inform decisions very broadly, it is important to understand how different policies could affect the outcomes of adaptation decision-making and the consequences of national or international mitigation actions. Performance-oriented indicators could be useful for informing adaptation options given budget restrictions or other constraints and for making choices to reduce vulnerabilities or increase response capacity. For example, the Environmental Performance Index (EPI) was developed based on feedback in part because the earlier Environmental Sustainability Index (ESI) did not focus on real policy levers that reflected what national governments could control. As a result, it may be useful to think about developing interactive policy tools to enable users to select alternative policies and observe the projected impact on the NCA indicators over time.

Indicators like the Consumer Price Index and the Unemployment Rate serve the dual purpose of providing integrated, easy-to-understand measures of the state of the economy and providing quantitative benchmarks that can be directly incorporated into decision algorithms, such as increases in salaries or reimbursement levels. These indicators are widely used by legislative and executive branch leaders and agencies, state and local governments, private firms, and private citizens to support decision-making. Designing analogous indicators relevant to climate impacts and adaptation could lead to their wider acceptance and use in policy and decision-making.

6.4.3 ABILITY TO ASSESS OVERALL SYSTEM STATE AND DYNAMICS, AGAINST WHICH THE BEHAVIOR OF INDIVIDUAL INDICATORS AND VARIABLES CAN BE JUDGED

Both the Earth's climate and human society are complex, interconnected systems that may exhibit heterogeneous responses to changing conditions, e.g., due to built-in lags, thresholds, and feedback mechanisms. Thus, it is very likely that some aspects of climate or of associated impacts will not behave in ways that seem consistent with other parts of the system, at least for some periods of time. For example, in the IPCC Fourth Assessment Report analysis of observed climate impacts, 84% of cells had observed climate impacts consistent with regional warming over at least two decades, but 13% had impacts that were not consistent (IPCC, 2007, Table 1.12). To scientists, these types of statistics are no surprise as they reflect the complexity of the systems in question and the likelihood of shifts in the spatial and temporal patterns of climate changes and societal responses. Policy makers and the public, in contrast, may have difficulty understanding the statistical details, which could lead to misunderstandings. However, a potential benefit of a structured set of indicators is that unusual changes (or lack of change) in a particular parameter, region, and/or time period can be assessed in the context of a more comprehensive view of the overall system. Such anomalies might only reflect internal variability or shifts in the systems, or could be harbingers of major changes in the entire system to come.

6.4.4 CONSISTENT INDICATOR FRAMEWORK ACROSS SOCIETAL, ECOLOGICAL, AND PHYSICAL TOPICS

The NCA will be developing an overarching indicator framework that includes indicators drawn from the societal, ecological, and physical indicator topic areas. An overarching indicator framework does not imply that there will be a single indicator that integrates everything, or hundreds of indicators that encompass anything related to climate. Instead, the goal is to provide a relatively small suite of indicators as described in Section 2. Ideally, these indicators would complement each other in a consistent manner. For example, it would be desirable for societal indicators of the impacts of extreme climate events to use definitions and baseline data compatible with the selected physical indicators of extreme events.

Though it is important to have a consistent indicator framework, there can still be flexibility in the presentation and approach for each of the component indicators. For example, it may be sensible to use a small basket of indicators to characterize different important dimensions of climate change based on multiple disaggregated climate variables. In contrast, to address whether the U.S. is successfully adapting to a changing climate, it might be useful to create a composite index that integrates economic measures for climate-sensitive sectors, e.g., a "Consumer Price Index-Climate" to track price changes associated with climate variability and change. Finally, to assess the vulnerability of populations to climate change it may be effective to present this information in a map format on a regional, state, or local scale to highlight patterns of vulnerability for the affected U.S. citizens. This has the benefit of presenting the information at the most appropriate scale, but it may not provide a means to easily aggregate the results, in non-visual format, at the national scale without losing key site-specific information.

6.5 Maintaining the Indicators

6.5.1 AVAILABILITY AND LONGEVITY OF KEY INPUT DATA

To develop and maintain indicators at the appropriate scale, data are essential. The NCA is developing an indicator framework with the goal to assess both short-term and long-term changes for climate sensitive societal impacts, adaptation, and vulnerability. Thus, in order for an indicator to have longevity, it is essential to identify long-term data sources for each indicator that can provide updates at appropriate intervals. This requires access to relevant data sources, a commitment to continue data collection and maintain documentation, and leveraging of the existing efforts of public and private partners.

After the implementation of the indicator system, there needs to be a process to adaptively assess the effectiveness of the indicators, update the indicators given better information, and add new indicators as priorities change over time. It is important to initially choose indicators that have staying power, while recognizing that as societal needs change and our understanding of the climate improves it is necessary to systematically evaluate the effectiveness of individual indicators for meeting the goal of the NCA indicator system.

6.5.2 RESOURCE AND TIME CONSTRAINTS

A key issue in maintaining indicators is to assess how each of the indicators will be deployed in decision-relevant timeframe and within a constrained budget environment. Leveraging the efforts of federal agencies and other NCA partners is essential. Ideally, candidate indicators would be maintained by different agencies or groups and incorporated with any necessary modifications into the NCA indicator system. An approach utilized by the Millennium Challenge Corporation (MCC) was to issue an open call for indicators and conduct a peer-review process to select indicators to be included in its basket. These indicators are maintained by the contributing organizations.

7 REFERENCES

Cutter, S., C.G. Burton, and C.T. Emrich. 2010. Disaster resilience indicators for benchmarking baseline conditions. *Journal of Homeland Security and Emergency Management* **7**: Article 51.

Eriksen, S., and R. Kelly. 2007. Developing credible vulnerability indicators for climate change adaptation policy assessment. *Mitigation and Adaptation Strategies for Global Change* **12**:495-524.

Florida, R., T. Gulden & C. Mellander. 2008. The rise of the mega-region. *Cambridge Journal of Regions, Economy and Society*, **1**:459-476.

IPCC. 2007. *Contribution of Working Group II to the Fourth Assessment Report of the Intergovernmental Panel on Climate Change*, 2007. M.L. Parry, O.F. Canziani, J.P. Palutikof, P.J. van der Linden and C.E. Hanson (eds). Cambridge University Press, Cambridge, United Kingdom and New York, NY, USA. 976 pp.

Keeney, R.L. and R.S. Gregory. 2005. Selecting attributes to measure the achievement of objectives. *Operations Research*. **53**:1-11.

Millennium Challenge Corporation (MCC). 2011. Guide to the MCC Indicators and Selection Process. 56pp. <http://www.mcc.gov/documents/reports/reference-2010001040503-_fy11guidetotheindicators.pdf>

Mills and Ebi. 2011. *National Climate Assessment Vulnerability Workshop Report*. Draft. Washington, DC: USGCRP National Climate Assessment.

Morin, A. 2005. The Canadian Water Sustainability Index (CWSI). In *PRI Working Paper Series no. 11.* Ottawa: Policy Research Initiative.

Moss, R., A.L. Brenkert, and E.L. Malone. 2001. *Vulnerability to Climate Change: A Quantitative Approach*. PNNL-SA-33642. Pacific Northwest National Laboratory.

NASA. 2007. *Earth Science and Applications from Space: National Imperatives for the Next Decade and Beyond.* Washington DC: National Academies Press.

NRC. 1999. *Our Common Journey: a Transition toward Sustainability*. Washington DC: National Academies Press.

NRC. 2010a. *America's Climate Choices: Adapting to the Impacts of Climate Change*. Washington, D.C.: National Research Council, The National Academies Press.

NRC. 2010b. *Monitoring Climate Change Impacts: Metrics at the Intersection of the Human and Earth Systems*. Washington, DC: National Research Council, The National Academies Press.

NRC. 2006. *Drawing Louisiana's New Map: Addressing Land Loss in Coastal Louisiana*. Washington, D.C.: National Research Council, The National Academies Press.

Schepelmann, P., Y. Goossens and A. Makipaa. 2010. Towards sustainable development: alternatives to GDP for measuring progress. Wuppertal Institute for Climate, Environment and Energy.

Sherrieb, K., F. Norris, and S. Galea. 2010. Measuring capacities for community resilience, *Social Indicators Research*, **99**:227-247.

Part 3: Societal Indicators Bibliography

Prepared by: Sandra R. Baptista

Abdallah, S., S. Mahony, N. Marks, J. Michaelson, C. Seaford, L. Stoll, and S. Thompson. 2011. Measuring our Progress: The Power of Well-being. London: The Centre for Well-being, New Economics Foundation.

Abdallah, S., S. Thompson, J. Michaelson, N. Marks, and N. Steuer. 2009. The (Un)Happy Planet Index 2.0: Why good lives don't have to cost the Earth. London: New Economics Foundation.

Adger, W. N., N. W. Arnell, and E. L. Tompkins. 2005. Successful adaptation to climate change across scales. Global Environmental Change 15 (2):77-86.

Adger, W. N., N. Brooks, G. Bentham, M. Agnew, and S. Eriksen. 2004. New indicators of vulnerability and adaptive capacity. In Technical Report 7. Norwich, UK: Tyndall Centre for Climate Change Research.

Adger, W. N., T. P. Hughes, C. Folke, S. R. Carpenter, and J. Rockstrom. 2005. Social-ecological resilience to coastal disasters. Science 309:1036-1039.

Ahlheim, M., and O. Frör. 2005. Constructing a Preference-Oriented Index of Environmental Quality. In Advances in Public Economics: Utility, Choice and Welfare, eds. U. Schmidt and S. Traub, 151-172: Springer US.

Alberti, M., and J. D. Parker. 1991. Indices of environmental quality: the search for credible measures. Environmental Impact Assessment Review 11 (2):95-101.

Albuquerque Indicators Progress Commission (AIPC). 2008. 2008 Albuquerque Progress Report. Albuquerque, NM.

Alder, J., S. Cullis-Suzuki, V. Karpouzi, K. Kaschner, S. Mondoux, W. Swartz, P. Trujillo, R. Watson, and D. Pauly. 2010. Aggregate performance in managing marine ecosystems of 53 maritime countries. Marine Policy 34 (3):468-476.

Alessa, L., and F. S. Chapin III. 2008. Anthropogenic biomes: a key contribution to earth-system science. Trends in Ecology & Evolution 23 (10):529-531.

Alessa, L., A. Kliskey, R. Lammers, C. Arp, D. White, L. Hinzman, and R. Busey. 2008. The Arctic Water Resource Vulnerability Index: an integrated assessment tool for community resilience and vulnerability with respect to freshwater. Environmental Management 42 (3):523-541.

Allison, E. H., A. L. Perry, M.-C. Badjeck, W. N. Adger, K. Brown, D. Conway, A. S. Halls, G. M. Pilling, J. D. Reynolds, N. L. Andrew, and N. K. Dulvy. 2009. Vulnerability of national economies to the impacts of climate change on fisheries. Fish and Fisheries 10 (2):173-196.

Anand, S., and A. Sen. 2000. The income component of the Human Development Index. Journal of Human Development 1 (1):83-106.

Andrews, F. M. 1989. The evolution of a movement. Journal of Public Policy 9 (4):401-405.

Armour, L. 2011. The puzzles and paradoxes of human need: an introduction. International Journal of Social Economics 38 (3):180-191.

Ashton, A. D., J. P. Donnelly, and R. L. Evans. 2008. A discussion of the potential impacts of climate change on the shorelines of the Northeastern USA. Mitigation and Adaptation Strategies for Global Change 13 (7):719-743.

Aubauer, H. P. 2011. Development of Ecological Footprint to an essential economic and political tool. Sustainability 3 (4):649-665.

Azar, D., and D. Rain. 2007. Identifying population vulnerable to hydrological hazards in San Juan, Puerto Rico. GeoJournal 69 (1-2):23-43.

Bagolin, I. P., and F. V. Comim. 2008. Human Development Index (HDI) and its family of indexes: an evolving critical review. Revista de Economia 34 (2):7-28.

Bagstad, K. J., and M. Ceroni. 2007. Opportunities and challenges in applying the Genuine Progress Indicator and Index of Sustainable Economic Welfare at local scales. International Journal of Environment, Workplace and Employment 3 (2):132-153.

————. 2008. The Genuine Progress Indicator: a new measure of economic development for the Northern Forest. Adirondack Journal of Environmental Studies 15 (1):(available online).

Bahadur, A. V., M. Ibrahim, and T. Tanner. 2010. The resilience renaissance? Unpacking of resilience for tackling climate change and disasters. Strengthening Climate Resilience, Discussion Paper 1. Brighton, UK: Institute of Development Studies.

Balica, S., and N. G. Wright. 2009. A network of knowledge on applying an indicator-based methodology for minimizing flood vulnerability. Hydrological Processes 23 (20):2983-2986.

————. 2010. Reducing the complexity of the flood vulnerability index. Environmental Hazards 9 (4):321-339.

Balica, S. F. 2007. Development and Application of Flood Vulnerability Indices for Various Spatial Scales, Water Science and Engineering, UNESCO-IHE, Delft.

Balica, S. F., N. Douben, and N. G. Wright. 2009. Flood vulnerability indices at varying spatial scales. Water Science & Technology 60 (10):2571-2580.

Bandura, R. 2008. A Survey of Composite Indices Measuring Country Performance: 2008 Update. In A UNDP/ODS Working Paper. New York: Office of Development Studies, United Nations Development Programme.

Barnett, J., S. Lambert, and I. Fry. 2008. The hazards of indicators: insights from the Environmental Vulnerability Index. Annals of the Association of American Geographers 98 (1):102-119.

Bauer, A., J. Feichtinger, and R. Steurer. 2011. The governance of climate change adaptation in ten OECD countries: challenges and approaches: Institute of Forest, Environmental, and Natural Resource Policy, University of Natural Resources and Applied Life Sciences, Vienna (BOKU).

Beça, P., and R. Santos. 2010. Measuring sustainable welfare: A new approach to the ISEW. Ecological Economics 69 (4):810-819.

Beeferman, L. W. 2002. The Asset Index: Assessing the Progress of States in Promoting Economic Security and Opportunity. Waltham, MA: Asset Development Institute, Center on Hunger and Poverty, The Heller Graduate School for Social Policy and Management, Brandeis University.

Beier, C. M., T. M. Patterson, and F. S. Chapin. 2008. Ecosystem services and emergent vulnerability in managed ecosystems: a geospatial decision-support tool. Ecosystems 11 (6):923-938.

Bekkedal, M. Y. V., K. M. Malecki, M. A. Werner, and H. A. Anderson. 2008. Using a partnership barometer to evaluate Environmental Public Health Tracking activities. Journal of Public Health Management & Practice 14 (6):592-595.

Bell, M. L., L. A. Cifuentes, D. L. Davis, E. Cushing, A. G. Telles, and N. Gouveia. 2011. Environmental health indicators and a case study of air pollution in Latin American cities. Environmental Research 111 (1):57-66.

Bell, S., and S. Morse. 2008 (2nd ed.). Sustainability Indicators: Measuring the Immeasurable? Second ed. London: Earthscan.

———. 2010. Rich pictures: a means to explore the 'sustainable mind'? Sustainable Development (published online).

———. 2011. An analysis of the factors influencing the use of indicators in the European Union. Local Environment: The International Journal of Justice and Sustainability 16 (3):281-302.

Bellani, L., and C. D'Ambrosio. 2010. Deprivation, social exclusion and subjective well-being. Social Indicators Research (published online).

Bellefleur, D., A. Bagnall, M. Mommaerts, and E. Plagman. 2010. Evaluation of the U.S. Government Millennium Challenge Corporation "Investing In People" Indicators. Prepared for the Millennium Challenge Corporation: Robert M. La Follette School of Public Affairs, University of Wisconsin-Madison.

Berik, G., and E. Gaddis. 2011. The Utah Genuine Progress Indicator (GPI), 1990 to 2007: A Report to the People of Utah. Available at: www.utahpop.org/gpi.html.

Bilgin, M. 2011. The PEARL model of sustainable development. Social Indicators Research (published online).

Binder, L., J. Barcelos, D. Booth, M. Darzen, M. Elsner, R. Fenske, T. Graham, A. Hamlet, J. Hodges-Howell, J. Jackson, C. Karr, P. Keys, J. Littell, N. Mantua, J. Marlow, D. McKenzie, M. Robinson-Dorn, E. Rosenberg, C. Stöckle, and J. Vano. 2010. Preparing for climate change in Washington State. Climatic Change 102 (1):351-376.

Birkmann, J. 2007. Risk and vulnerability indicators at different scales: Applicability, usefulness and policy implications. Environmental Hazards 7 (1):20-31.

Bjarnadottir, S., Y. Li, and M. Stewart. 2011. Social vulnerability index for coastal communities at risk to hurricane hazard and a changing climate. Natural Hazards (published online).

Blanc, I., D. Friot, M. Margni, and O. Jolliet. 2008. Towards a new index for environmental sustainability based on a DALY weighting approach. Sustainable Development 16 (4):251-260.

Bleys, B. 2008. Proposed changes to the Index of Sustainable Economic Welfare: An application to Belgium. Ecological Economics 64 (4):741-751.

BLS. 2010. Experimental Consumer Price Index for Americans 62 Years of Age and Older, 1998-2009. Washington, D.C.: Bureau of Labor Statistics.

Böhringer, C., and P. E. P. Jochem. 2007. Measuring the immeasurable — A survey of sustainability indices. Ecological Economics 63 (1):1-8.

Boicourt, K., and Z. P. Johnson. 2010. Comprehensive Strategy for Reducing Maryland's Vulnerability to Climate Change, Phase II: Building societal, economic, and ecological resilience. Report of the Maryland Commission on Climate Change, Adaptation and Response and Scientific and Technical Working Groups: University of Maryland Center for Environmental Science, Cambridge, Maryland and Maryland Department of Natural Resources, Annapolis, Maryland.

Booysen, F. 2002. An overview and evaluation of composite indices of development. Social Indicators Research 59 (2):115-151.

Borden, K. A., M. C. Schmidtlein, C. T. Emrich, W. W. Piegorsch, and S. L. Cutter. 2007. Vulnerability of U.S. Cities to Environmental Hazards. Journal of Homeland Security and Emergency Management 4 (2):Article 5.

Bossel, H. 1999. Indicators for sustainable development: theory, method, applications. A report to the Balaton Group: International Institute for Sustainable Development.

Boulanger, P.-M. 2007. Political uses of social indicators: overview and application to sustainable development indicators. Int. J. Sustainable Development 10 (1/2):14-32.

Bowen, R. E., and C. Riley. 2003. Socio-economic indicators and integrated coastal management. Ocean & Coastal Management 46 (3-4):299-312.

Bradshaw, J., and D. Richardson. 2009. An index of child well-being in Europe. Child Indicators Research 2 (3):319-351.

Brooks, N., and W. N. Adger. 2003. Country level risk measures of climate-related natural disasters and implications for adaptation to climate change. Norwich, UK: Tyndall Centre for Climate Change Research.

Brugmann, J. 1997. Is there a method in our measurement? The use of indicators in local sustainable development planning. Local Environment: The International Journal of Justice and Sustainability 2 (1):59-72.

Brumbaugh-Smith, J., H. Gross, N. Wollman, and B. Yoder. 2008. NIVAH: a composite index measuring violence and harm in the U.S. Social Indicators Research 85 (3):351-387.

Bulmer, M. 1989. Problems of theory and measurement. Journal of Public Policy 9 (4):407-412.

Burton, C. G. 2010. Social vulnerability and hurricane impact modeling. Natural Hazards Review 11 (2):58-68.

Buys, P., U. Deichmann, C. Meisner, T. T. That, and D. Wheeler. 2009. Country stakes in climate change negotiations: two dimensions of vulnerability. Climate Policy 9 (3):288-305.

Callebaut, W. 1978. Social indicators research and the theory of collective action. Pholosophica 21 (1):159-197.

Campbell, A., V. Kapos, A. Chenery, S. I. Kahn, M. Rashid, J. Scharlemann, and B. Dickson. 2009. The linkages between biodiversity and climate change adaptation: a review of the recent scientific literature: United Nations Environment Programme World Conservation Monitoring Centre (UNEP-WCMC).

Cardona, O. D. 2007. Indicators of disaster risk and risk management: program for Latin America and the Caribbean: summary report. Washington, D.C.: Inter-American Development Bank (IDB), Sustainable Development Department, Environment Division.

Cardona, O. D., and M. L. Carreño. 2011. Updating the Indicators of Disaster Risk and Risk Management for the Americas. IDRiM (Journal of Integrated Disaster Risk Management) 1 (1):(published online).

Cardona, O. D., M. G. Ordaz, M. C. Marulanda, M. L. Carreño, and A. H. Barbat. 2010. Disaster risk from a macroeconomic perspective: a metric for fiscal vulnerability evaluation. Disasters 34 (4):1064-1083.

Carmichael, N., and R. Parke. 1974. Information services for social indicators research. Special Libraries 55 (5-6).

Carreño, M. L., O. D. Cardona, and A. H. Barbat. 2007. A disaster risk management performance index. Natural Hazards 41 (1):1-20.

Carvalho, J. F. d. 2011. Measuring economic performance, social progress and sustainability using an index. Renewable and Sustainable Energy Reviews 15 (2):1073-1079.

Castañeda, B. E. 1999. An index of sustainable economic welfare (ISEW) for Chile. Ecological Economics 28 (2):231-244.

Cendrero, A., E. Francés, D. D. Corral, J. L. Fermán, D. Fischer, L. D. Río, M. Camino, and A. López. 2003. Indicators and indices of environmental quality for sustainability assessment in coastal areas: application to case studies in Europe and the Americas. Journal of Coastal Research 19 (4):919-933.

Centre for Indigenous Environmental Resources (CIER), and A. Morin. 2006. The Canadian Water Sustainability Index (CWSI) Case Study Report (PRI Working Paper Series #028): Policy Research Initiative, Government of Canada.

Chakravarty, S. 2003. A generalized Human Development Index. Review of Development Economics 7 (1):99-114.

Chapin III, F. S., S. R. Carpenter, G. P. Kofinas, C. Folke, N. Abel, W. C. Clark, P. Olsson, D. M. S. Smith, B. Walker, O. R. Young, F. Berkes, R. Biggs, J. M. Grove, R. L. Naylor, E. Pinkerton, W. Steffen, and F. J. Swanson. 2010. Ecosystem stewardship: sustainability strategies for a rapidly changing planet. Trends in Ecology & Evolution 25 (4):241-249.

Charleston, A. E., A. Banerjee, and V. G. Carande-Kulis. 2008. Measuring success: the case for calculating the return on investment of Environmental Public Health Tracking. Journal of Public Health Management & Practice 14 (6):600-604.

Charleston, A. E., P. Wall, C. Kassinger, and P. O. Edwards. 2008. Implementing the Environmental Public Health Tracking Network: accomplishments, challenges, and directions. Journal of Public Health Management and Practice 14 (6):507-514.

Chaves, H. M. L., and S. Alipaz. 2007. An Integrated Indicator Based on Basin Hydrology, Environment, Life, and Policy: The Watershed Sustainability Index. Water Resources Management 21 (5):883-895.

Cherchye, L., W. Moesen, N. Rogge, and T. Van Puyenbroeck. 2007. An introduction to 'benefit of the doubt' composite indicators. Social Indicators Research 82 (1):111-145.

———. 2011. Constructing composite indicators with imprecise data: a proposal. Expert Systems with Applications 38 (9):10940-10949.
Cherchye, L., E. Ooghe, and T. Van Puyenbroeck. 2008. Robust human development rankings. Journal of Economic Inequality 6 (4):287-321.

Cho, D. I., T. Ogwang, and C. Opio. 2010. Simplifying the Water Poverty Index. Social Indicators Research 97 (2):257-267.

Christopherson, S., J. Michie, and P. Tyler. 2010. Regional resilience: theoretical and empirical perspectives. Cambridge Journal of Regions, Economy and Society 3 (1):3-10.

Clarke, M., and P. Lawn. 2008. Is measuring genuine progress at the sub-national level useful? Ecological Indicators 8 (5):573-581.

———. 2008. A policy analysis of Victoria's Genuine Progress Indictor. Journal of Socio-Economics 37 (2):864-879.

Cobb, C., T. Halstead, and J. Rowe. 1995. If the GDP is up, why is America down? The Atlantic Monthly 276 (4):59-78.

Cobb, C., and C. Rixford. 1998. Competing paradigms in the development of social and economic indicators. In Prepared for the Centre for the Study of Living Standards (CSLS) Conference on the State of Living Standards and the Quality of Life in Canada. Ottawa, Ontario.

Coelho, P., A. Mascarenhas, P. Vaz, A. Dores, and T. B. Ramos. 2010. A framework for regional sustainability assessment: developing indicators for a Portuguese region. Sustainable Development 18 (4):211-219.

Compton, R. A., D. C. Giedeman, and G. A. Hoover. 2011. Panel evidence on economic freedom and growth in the United States. European Journal of Political Economy 27 (3):423-435.

Connor, R. F., and K. Hiroki. 2005. Development of a method for assessing flood vulnerability. Water Science & Technology 51 (5):61-67.

Coombes, E. G., A. P. Jones, and W. J. Sutherland. 2009. The Implications of Climate Change on Coastal Visitor Numbers: A Regional Analysis. Journal of Coastal Research 25 (4):981-990.

Costanza, R., J. Erickson, K. Fligger, A. Adams, C. Adams, B. Altschuler, S. Balter, B. Fisher, J. Hike, J. Kelly, T. Kerr, M. McCauley, K. Montone, M. Rauch, K. Schmiedeskamp, D. Saxton, L. Sparacino, W. Tusinski, and L. Williams. 2004. Estimates of the Genuine Progress Indicator (GPI) for Vermont, Chittenden County and Burlington, from 1950 to 2000. Ecological Economics 51 (1-2):139-155.

Costanza, R., M. Hart, S. Posner, and J. Talberth. 2009. Beyond GDP: The Need for New Measures of Progress. In The Pardee Papers. Boston: The Frederick S. Pardee Center for the Study of the Longer-Range Future, Boston University.

Coulthard, S., D. Johnson, and J. A. McGregor. 2011. Poverty, sustainability and human wellbeing: A social wellbeing approach to the global fisheries crisis. Global Environmental Change 21 (2):453-463.

Coulton, C. J. 2008. Catalog of Administrative Data Sources for Neighborhood Indicators: A National Neighborhood Indicators Partnership Guide. Washington, D.C.: The Urban Institute.

Coulton, C. J., and R. L. Fischer. 2010. Using Early Childhood Wellbeing Indicators to Influence Local Policy and Services. In From Child Welfare to Child Well-being, eds. S. B. Kamerman, S. Phipps and A. Ben-Arieh, 101-116: Springer Netherlands.

Coulton, C. J., and J. E. Korbin. 2007. Indicators of child well-being through a neighborhood lens. Social Indicators Research 84 (3):349-361.

Coulton, C. J., J. E. Korbin, and J. McDonell. 2009. Editorial: Indicators of child well-being in the context of small areas. Child Indicators Research 2 (2):109-110.

Cranston, G. R., and G. P. Hammond. 2010. North and south: Regional footprints on the transition pathway towards a low carbon, global economy. Applied Energy 87 (9):2945-2951.

Cranston, G. R., G. P. Hammond, and R. C. Johnson. 2010. Ecological Debt: Exploring the Factors that Affect National Footprints. Journal of Environmental Policy & Planning 12 (2):121-140.

CTSIP. 2010. Central Texas Sustainability Indicators Project: 2009 Indicators Report. Austin, TX: CTSIP.

Cummins, R. A. 1996. The domains of life satisfaction: An attempt to order chaos. Social Indicators Research 38 (3):303-328.

Cummins, R. A., M. P. Mccabe, Y. Romeo, and E. Gullone. 1994. Validity Studies the Comprehensive Quality of Life Scale (Comqol): Instrument Development and Psychometric Evaluation on College Staff and Students Educational and Psychological Measurement 54 (2):372-382.

Cust, J. 2009. Using intermediate indicators: lessons for climate policy. Climate Policy 9 (5):450-463.

Cutter, S., L. Barnes, M. Berry, C. Burton, E. Evans, E. Tate, and J. Webb. 2008. A place-based model for understanding community resilience to natural disasters. Global Environmental Change 18 (4):598-606.

Cutter, S. L. 2010. Social science perspectives on hazards and vulnerability science. Geophysical Hazards 1:17-30.

Cutter, S. L., B. J. Boruff, and W. L. Shirley. 2003. Social vulnerability to environmental hazards. Social Science Quarterly 84 (2):242-261.

Cutter, S. L., C. G. Burton, and C. T. Emrich. 2010. Disaster resilience indicators for benchmarking baseline conditions. Journal of Homeland Security and Emergency Management 7 (1):Article 51.

Cutter, S. L., C. T. Emrich, J. J. Webb, and D. Morath. 2009. Social vulnerability to climate variability hazards: a review of the literature: final report to Oxfam America: Hazards and Vulnerability Research Institute, University of South Carolina.

Cutter, S. L., and C. Finch. 2008. Temporal and spatial changes in social vulnerability to natural hazards. Proceedings of the National Academy of Sciences 105 (7):2301-2306.

Cutter, S. L., J. T. Mitchell, and M. S. Scott. 2000. Revealing the vulnerability of people and places: a case study of Georgetown County, South Carolina. Annals of the Association of American Geographers 90 (4):713-737.

D'Acci, L. 2010. Measuring well-being and progress. Social Indicators Research (published online 1 October 2010).

Dalcanale, F., D. Fontane, and J. Csapo. 2011. A general framework for a collaborative Water Quality Knowledge and Information Network. Environmental Management 47 (3):443-455.

Daly, H. E., and J. B. Cobb Jr. 1989. For the Common Good: Redirecting the Economy toward Community, the Environment, and a Sustainable Future. Boston: Beacon Press.
DARA. 2010. Climate Vulnerability Monitor 2010: The State of the Climate Crisis. Madrid: DARA and the Climate Vulnerable Forum.

Davern, M., and X. Chen. 2010. Piloting the Geographic Information System (GIS) methodology as an analytic tool for Subjective Wellbeing research. Applied Research in Quality of Life 5 (2).

Davidson, K. 2005. Towards an integrated sustainability indicator framework. International Journal of Environment, Workplace and Employment 1 (3-4):370-382.

Davidson, K. M. 2011. Reporting systems for sustainability: What are they measuring? Social Indicators Research 100 (2):351-365.

De Stefano, L. 2010. International initiatives for water policy assessment: a review. Water Resources Management 24 (11):2449-2466.

Debels, P., C. Szlafsztein, P. Aldunce, C. Neri, Y. Carvajal, M. Quintero-Angel, A. Celis, A. Bezanilla, and D. Martínez. 2009. IUPA: a tool for the evaluation of the general usefulness of practices for adaptation to climate change and variability. Natural Hazards 50 (2):211-233.

DESA. 2005. Understanding Knowledge Societies: In twenty questions and answers with the Index of Knowledge Societies. New York: United Nations Department of Economic and Social Affairs.

———. 2007. Indicators of Sustainable Development: Guidelines and Methodologies (3rd edition). New York: United Nations, Department of Economic and Social Affairs.

Deutsch, L., A. Jansson, M. Troell, P. Ronnback, C. Folke, and N. Kautsky. 2000. The "ecological footprint": communicating human dependence on nature's work. Ecological Economics 32 (3):351-355.

Dever, G. E. A. 1979. Social indicators, 1976: a critique. Social Indicators Research 6 (2):153-162.

Diener, E. 1995. A value based index for measuring national quality of life. Social Indicators Research 36 (2):107-127.

Dietz, S., and E. Neumayer. 2006. Some constructive criticisms of the Index of Sustainable Economic Welfare. In Sustainable Development Indicators in Ecological Economics, ed. P. Lawn, 186-206. Cheltenham, UK: Edward Elgar.

Dietz, T., E. A. Rosa, and R. York. 2009. Environmentally efficient well-being: rethinking sustainability as the relationship between human well-being and environmental impacts. Human Ecology Review 16 (1):114-123.

———. 2010. Environmentally efficient well-being: Is there a Kuznets curve? Applied Geography (in press).

Diewert, W. E., A. O. Nakamura, and L. I. Nakamura. 2009. The housing bubble and a new approach to accounting for housing in a CPI. Journal of Housing Economics 18 (3):156-171.

Dillard, M. 2010. Toward a measure of social-ecological resilience for human communities. In Shifting Shorelines: Adapting to the Future, The 22nd International Conference of The Coastal Society. Wilmington, NC.

Dluhy, M., and N. Swartz. 2006. Connecting knowledge and policy: the promise of community indicators in the United States. Social Indicators Research 79 (1):1-23.

Domínguez-Serrano, M., and F. Blancas. 2010. A gender wellbeing composite indicator: the best-worst global evaluation approach. Social Indicators Research 102 (3):477-496.

Downton, M. W., and R. A. Pielke. 2005. How Accurate are Disaster Loss Data? The Case of U.S. Flood Damage. Natural Hazards 35 (2):211-228.

Dragomirescu, H., and R. S. Sharma. 2009. Operationalising the Sustainable Knowledge Society Concept through a Multi-dimensional Scorecard. In Best Practices for the Knowledge Society. Knowledge, Learning, Development and Technology for All, eds. M. D. Lytras, P. Ordonez de Pablos, E. Damiani, D. Avison, A. Naeve and D. G. Horner, 328-337. Berlin Heidelberg: Springer.

Dreher, A. 2006. Does globalization affect growth? Evidence from a new Index of Globalization. Applied Economics 38 (10):1091-1110.

Dreher, A., N. Gaston, and P. Martens. 2008. Measuring Globalisation: Gauging its Consequences. New York: Springer.

Ebert, U., and H. Welsch. 2004. Meaningful environmental indices: a social choice approach. Journal of Environmental Economics and Management 47 (2):270-283.

Elmore, K., B. Flanagan, N. Jones, and J. Heitgerd. 2010. Leveraging geospatial data, technology, and methods for improving the health of communities: priorities and strategies from an expert panel convened by the CDC. Journal of Community Health 35 (2):165-171.

Engel, K. H., and M. L. Miller. 2009. State governance: leadership on climate change. In Agenda for a Sustainable America, ed. J. C. Dernbach, 441-456. Washington, D.C.: Environmental Law Institute.

England, R. W. 1998. Measurement of social well-being: alternatives to gross domestic product. Ecological Economics 25 (1):89-103.

Engle, N. L. 2011. Adaptive capacity and its assessment. Global Environmental Change 21 (2):647-656.

English, P. B., A. H. Sinclair, Z. Ross, H. Anderson, V. Boothe, C. Davis, K. Ebi, B. Kagey, K. Malecki, R. Shultz, and E. Simms. 2009. Environmental health indicators of climate change for the United States: findings from the State Environmental Health Indicator Collaborative. Environmental Health Perspectives 117 (11):1673-1681.

Environmental Protection Agency (EPA). 2010. Climate Change Indicators in the United States. Washington, D.C.: EPA.

Eriksen, S. H., and P. M. Kelly. 2007. Developing credible vulnerability indicators for climate adaptation policy assessment. Mitigation and Adaptation Strategies for Global Change 12 (4):495-524.

Esnard, A.-M., A. Sapat, and D. Mitsova. 2011. An index of relative displacement risk to hurricanes. Natural Hazards:(published online).

Esty, D. C., M. Levy, T. Srebotnjak, and A. de Sherbinin. 2005. 2005 Environmental Sustainability Index: Benchmarking National Environmental Stewardship. New Haven: Yale Center for Environmental Law & Policy.

European Environment Agency (EEA). 2005. EEA core set of indicators -- Guide. In Technical report No. 1/2005. Copenhagen, Denmark: European Environment Agency.

Ewers, R. M., and R. J. Smith. 2007. Choice of index determines the relationship between corruption and environmental sustainability. Ecology and Society 12 (1):r2.

Ewing, B., D. Moore, S. Goldfinger, A. Oursler, A. Reed, and M. Wackernagel. 2010. The Ecological Footprint Atlas 2010. Oakland, CA: Global Footprint Network.

Fath, B. D., and H. Cabezas. 2004. Exergy and Fisher Information as ecological indices. Ecological Modelling 174 (1-2):25-35.

Ferriss, A. L. 1975. National approaches to developing social indicators. Social Indicators Research 2 (1):81-92.

———. 1979. The U.S. Federal effort in developing social indicators. Social Indicators Research 6 (2):129-152.

———. 1988. The uses of social indicators. Social Forces 66 (3):601-617.

———. 1989. Whatever happened, indeed! Journal of Public Policy 9 (4):413-417.

———. 2000. The quality of life among U.S. States. Social Indicators Research 49 (1):1-23.

———. 2004. The quality of life concept in sociology. The American Sociologist 35 (3):37-51.

Fiala, N. 2008. Measuring sustainability: Why the ecological footprint is bad economics and bad environmental science. Ecological Economics 67 (4):519-525.

Flanagan, B. E., E. W. Gregory, E. J. Hallisey, J. L. Heitgerd, and B. Lewis. 2011. A Social Vulnerability Index for disaster management. Journal of Homeland Security and Emergency Management 8 (1):Article 3.

Florida, R., T. Gulden, and C. Mellander. 2008. The rise of the mega-region. Cambridge Journal of Regions, Economy and Society 1:459-476.

Fontalvo-Herazo, M. L., M. Glaser, and A. Lobato-Ribeiro. 2007. A method for the participatory design of an indicator system as a tool for local coastal management. Ocean & Coastal Management 50 (10):779-795.

Ford, J. D., E. C. H. Keskitalo, T. Smith, T. Pearce, L. Berrang-Ford, F. Duerden, and B. Smit. 2010. Case study and analogue methodologies in climate change vulnerability research. Wiley Interdisciplinary Reviews: Climate Change 1 (3):374-392.

Fraser, E. D. G., A. J. Dougill, W. E. Mabee, M. Reed, and P. McAlpine. 2006. Bottom up and top down: analysis of participatory processes for sustainability indicator identification as a pathway to community empowerment and sustainable environmental management. Journal of Environmental Management 78 (2):114-127.

Frecker, K. 2005. Beyond GDP: enabling democracy with better measures of social well-being: Trudeau Centre for Peace and Conflict Studies, University of Toronto.

Fredericks, S. E. 2011. Monitoring Environmental Justice. Environmental Justice 4 (1):63-69.

Frumhoff, P. C., J. J. McCarthy, J. Melillo, S. C. Moser, and D. J. Wuebbles. 2007. Confronting Climate Change in the U.S. Northeast: Science, Impacts, and Solutions. Synthesis report of the Northeast Climate Impacts Assessment (NECIA). Cambridge, MA: Union of Concerned Scientists.

Frumkin, H., J. Hess, G. Luber, J. Malilay, and M. McGeehin. 2008. Climate change: the public health response. American Journal of Public Health 98 (3):435-445.

Füssel, H.-M. 2009. Review and quantitative analysis of indices of climate change exposure, adaptive capacity, sensitivity, and impacts (Background note to the World Development Report 2010). Washington, D.C.: World Bank.

———. 2010. How inequitable is the global distribution of responsibility, capability, and vulnerability to climate change: a comprehensive indicator-based assessment. Global Environmental Change 20 (4):597-611.

Gahin, R., V. Veleva, and M. Hart. 2003. Do indicators help create sustainable communities? Local Environment: The International Journal of Justice and Sustainability 8 (6):661-666.

Gallego-Álvarez, I., L. Rodríguez-Domínguez, and I.-M. García-Sánchez. 2010. Are determining factors of municipal E-government common to a worldwide municipal view? An intra-country comparison. Government Information Quarterly 27 (4):423-430.

Gallopín, G. C. 1996. Environmental and sustainability indicators and the concept of situational indicators: a systems approach. Environmental Modeling and Assessment 1 (3):101-117.

———. 1997. Indicators and their use: information for decision-making. In Sustainability indicators: a report on the project on sustainability indicators of sustainable development, 13-27. Chichester: John Wiley.

Ganning, J. P., and C. G. Flint. 2010. Constructing a community-level amenity index. Society & Natural Resources: An International Journal 23 (12):1253-1258.

Gasparatos, A. 2010. Embedded value systems in sustainability assessment tools and their implications. Journal of Environmental Management 91 (8):1613-1622.

Gasparatos, A., M. El-Haram, and M. Horner. 2008. A critical review of reductionist approaches for assessing the progress towards sustainability. Environmental Impact Assessment Review 28 (4-5):286-311.

Gasper, D. 2010. Understanding the diversity of conceptions of well-being and quality of life. Journal of Socio-Economics 39 (3):351-360.

Gasteyer, S., and C. B. Flora. 1999. Social indicators: an annotated bibliography on trends, sources and development, 1960-1998.

Giannetti, B. F., S. H. Bonilla, C. C. Silva, and C. M. V. B. Almeida. 2009. The reliability of experts' opinions in constructing a composite environmental index: The case of ESI 2005. Journal of Environmental Management 90 (8):2448-2459.

Gil, S., and J. Sleszynski. 2003. An index of sustainable economic welfare for Poland. Sustainable Development 11 (1):47-55.

Giljum, S., E. Burger, F. Hinterberger, S. Lutter, and M. Bruckner. 2011. A comprehensive set of resource use indicators from the micro to the macro level. Resources, Conservation and Recycling 55 (3):300-308.

Giuliani, G., and P. Peduzzi. 2011. The PREVIEW Global Risk Data Platform: a geoportal to serve and share global data on risk to natural hazards. Natural Hazards and Earth System Sciences 11:53-66.

Gober, P. 2010. Desert urbanization and the challenges of water sustainability. Current Opinion in Environmental Sustainability 2 (3):144-150.

Gomez-Limon, J. A., and G. Sanchez-Fernandez. 2010. Empirical evaluation of agricultural sustainability using composite indicators. Ecological Economics 69 (5):1062-1075.

Gough, A. D., J. L. Innes, and S. D. Allen. 2008. Development of common indicators of sustainable forest management. Ecological Indicators 8 (5):425-430.

Graymore, M. L. M., N. G. Sipe, and R. E. Rickson. 2008. Regional sustainability: How useful are current tools of sustainability assessment at the regional scale? Ecological Economics 67 (3):362-372.

Greenlees, J. S., and R. B. McClelland. 2008. Addressing misconceptions about the Consumer Price Index. Monthly Labor Review August 2008:3-19.

Grimm, M., K. Harttgen, S. Klasen, and M. Misselhorn. 2008. A Human Development Index by Income Groups. World Development 36 (12):2527-2546.

Grimm, M., K. Harttgen, S. Klasen, M. Misselhorn, T. Munzi, and T. Smeeding. 2010. Inequality in Human Development: An Empirical Assessment of 32 Countries. Social Indicators Research 97 (2):191-211.

Grubesic, T. H., and T. C. Matisziw. 2011. A typological framework for categorizing infrastructure vulnerability. GeoJournal:(published online 03 May 2011).

Grupp, H., and T. Schubert. 2010. Review and new evidence on composite innovation indicators for evaluating national performance. Research Policy 39 (1):67-78.

Guhathakurta, S., and P. Gober. 2007. The impact of the Phoenix urban heat island on residential water use. Journal of the American Planning Association 73 (3):317-329.

Guy, G. B., and C. J. Kibert. 1998. Developing indicators of sustainability: US experience. Building Research & Information 26 (1):39-45.

Hagerty, M. R., R. A. Cummins, A. L. Ferriss, K. Land, A. C. Michalos, M. Peterson, A. Sharpe, J. Sirgy, and J. Vogel. 2001. Quality of life indexes for national policy: review and agenda for research. Social Indicators Research 55 (1):1-96.

Hagerty, M. R., and K. C. Land. 2007. Constructing summary indices of quality of life: a model for the effect of heterogeneous importance weights. Sociological Methods & Research 35 (4):455-496.

Hamilton, C. 1999. The genuine progress indicator methodological developments and results from Australia. Ecological Economics 30 (1):13-28.

Hanley, N., I. Moffatt, R. Faichney, and M. Wilson. 1999. Measuring sustainability: A time series of alternative indicators for Scotland. Ecological Economics 28 (1):55-73.

Harger, J. R. E., and F.-M. Meyer. 1996. Definition of indicators for environmentally sustainable development. Chemosphere 33 (9):1749-1775.

Hassink, R. 2010. Regional resilience: a promising concept to explain differences in regional economic adaptability? Cambridge Journal of Regions, Economy and Society 3 (1):45-58.

Hastings, D. A. 2008. Describing the human condition – from human development to human security: an environmental remote sensing and GIS approach. In GIS-IDEAS 2008 Conference. Hanoi, Vietnam.
———. 2009. From Human Development to Human Security: A Prototype Human Security Index. UNESCAP Working Paper WP/09/03, Macroeconomic Policy and Development Division. Bangkok, Thailand: United Nations Economic and Social Commission for Asia and the Pacific.

————. 2009. Filling Gaps in the Human Development Index: Findings for Asia and the Pacific. UNESCAP Working Paper WP/09/02, Macroeconomic Policy and Development Division. Bangkok, Thailand: United Nations Economic and Social Commission for Asia and the Pacific.

————. 2010. The global human security index: Can disaggregations help us to forge progress? . In Shifting Shorelines: Adapting to the Future,The 22nd International Conference of The Coastal Society. Wilmington, North Carolina.

————. 2011. The Human Security Index: Potential Roles for the Environmental and Earth Observation Communities. Earthzine (available online).

Hayati, D., Z. Ranjbar, and E. Karami. 2011. Measuring agricultural sustainability. In Biodiversity, Biofuels, Agroforestry and Conservation Agriculture, ed. E. Lichtfouse, 73-100: Springer Netherlands.

Heink, U., and I. Kowarik. 2010. What are indicators? On the definition of indicators in ecology and environmental planning. Ecological Indicators 10 (3):584-593.

Hezri, A. A., and S. R. Dovers. 2006. Sustainability indicators, policy and governance: Issues for ecological economics. Ecological Economics 60 (1):86-99.

————. 2009. Australia's Indicator-Based Sustainability Assessments and Public Policy. Australian Journal of Public Administration 68 (3):303-318.

Hinkel, J. 2011. "Indicators of vulnerability and adaptive capacity": Towards a clarification of the science–policy interface. Global Environmental Change 21 (1):198-208.

Holand, I. S., P. Lujala, and J. K. Rød. 2011. Social vulnerability assessment for Norway: A quantitative approach. Norsk Geografisk Tidsskrift - Norwegian Journal of Geography 65 (1):1-17.

Holden, M. 2006. Urban indicators and the integrative ideals of cities. Cities 23 (3):170-183.

————. 2009. Community interests and indicator system success. Social Indicators Research 92 (3):429-448.

Holman, N. 2009. Incorporating local sustainability indicators into structures of local governance: a review of the literature. Local Environment: The International Journal of Justice and Sustainability 14 (4):365-375.

Hopton, M. E., H. Cabezas, D. Campbell, T. Eason, A. S. Garmestani, M. T. Heberling, A. T. Karunanithi, J. J. Templeton, D. White, and M. Zanowick. 2010. Development of a multidisciplinary approach to assess regional sustainability International Journal of Sustainable Development & World Ecology 17 (1):48-56.

Horn, R. V. 1977. Government take-over of the social indicators movement? . Social Indicators Research 5 (1-4):365-367.

————. 1980. Social indicators: meaning, methods and applications. International Journal of Social Economics 7 (8):419-460.

Hovelsrud, G. K., J. L. White, M. Andrachuk, and B. Smit. 2010. Community Adaptation and Vulnerability Integrated. In Community Adaptation and Vulnerability in Arctic Regions, eds. G. K. Hovelsrud and B. Smit, 335-348: Springer Netherlands.
Howe, P. D. 2011. Hurricane preparedness as anticipatory adaptation: a case study of community businesses. Global Environmental Change 21 (2):711-720.

Howell, E. M., K. L. S. Pettit, B. A. Ormond, and G. T. Kingsley. 2003. Using the National Neighborhood Indicators Partnership to improve public health. Journal of Public Health Management & Practice 9 (3):235-242.

Hur, Y., and R. Testerman. 2010. An index of child well-being at a local level in the U.S.: the case of North Carolina counties. Child Indicators Research (In press).

Innes, J. E. 1975. Social Indicators and Public Policy: Interactive Processes of Design and Application. Amsterdam: Elsevier Scientific Publishing Co.

———. 1989. Disappointments and legacies of social indicators. Journal of Public Policy 9 (4):429-432.

———. 1990. Knowledge and Public Policy: The Search for Meaningful Indicators. New Brunswick, NJ: Rutgers Transaction Books.

———. 2010. Planning with Complexity: An Introduction to Collaborative Rationality for Public Policy. Oxford: Routledge/Taylor Francis.

Innes, J. E., D. E. Booher, and S. Di Vittorio. 2011. Strategies for megaregion governance: collaborative dialogue, networks, and self-organization. Journal of the American Planning Association 77 (1):55-67.

Inter-American Development Bank (IDB). The Prevalent Vulnerability Index (PVI): IDB.

———. The Disaster Deficit Index (DDI): IDB.

———. The Local Disaster Index (LDI): IDB.

Intergovernmental Panel on Climate Change (IPCC). 2007. Contribution of Working Group II to the Fourth Assessment Report of the Intergovernmental Panel on Climate Change, eds. M. L. Parry, O. F. Canziani, J. P. Palutikof, P. J. v. d. Linden and C. E. Hanson, 976pp. Cambridge, United Kingdom and New York, NY, USA.

Jacob, K., R. Black, R. Horton, D. Bader, and M. O'Grady. 2010. Chapter 7: Indicators and monitoring. Annals of the New York Academy of Sciences 1196:127-141.

Jalonne White-Newsome, M. S. O. N., Carina Gronlund, Tenaya M. Sunbury, Shannon J. Brines, Edith Parker, Daniel G. Brown, Richard B. Rood, Zorimar Rivera. 2009. Climate change, heat waves, and environmental justice: advancing knowledge and action. Environmental Justice 2 (4):197-205.

Johnston, D. F. 1979. Discussion: problems in social indicator development. Social Indicators Research 6 (2):207-213.

———. 1989. Some reflections on the United States. Journal of Public Policy 9 (4):433-436.

Jol, A., M. Erhard, F. Raes, J. V. Minnen, R. Swart, P. Sastny, R. V. Dingenen, T. Voigt, G. Sander, T. Christiansen, L. Feyen, P. Kristensen, M. Harley, A. Jones, L. Montanarella, V. Stolbovoy, A.-R. Gentile, C. Lavalle, F. Micale, C. Lazar, C. Conte, G. Genovese, A. Camia, T. Durrant, G. Amatulli, R. Hierderer, B. Menne, S. Isoard, P. Watkiss, and J. B. Cano. 2008. Impacts of Europe's changing climate – 2008 indicator-based assessment. In EEA Report No 4/2008. JRC Reference Report No JRC47756: EEA, JRC, WHO.

Jones, J. P. G., B. Collen, G. Atkinson, P. W. J. Baxter, P. Bubb, J. B. Illian, T. E. Katzner, A. Keane, J. Loh, E. McDonald-Madden, E. Nicholson, H. M. Pereira, H. P. Possingham, A. S. Pullin, A. S. L. Rodrigues, V. Ruiz-Gutierrez, S. Matthew, and E. J. Milner-Gulland. 2011. The why, what, and how of global biodiversity indicators beyond the 2010 target. Conservation Biology 25 (3):450-457.

Jordan, S. J., S. E. Hayes, D. Yoskowitz, L. M. Smith, J. K. Summers, M. Russell, and W. H. Benson. 2010. Accounting for natural resources and environmental sustainability: linking ecosystem services to human well-being. Environmental Science & Technology 44 (5):1530-1536.

Jorgenson, A. K., and T. J. Burns. 2007. The political-economic causes of change in the ecological footprints of nations, 1991-2001: A quantitative investigation. Social Science Research 36 (2):834-853.

Jorgenson, A. K., and B. Clark. 2011. Societies consuming nature: A panel study of the ecological footprints of nations, 1960-2003. Social Science Research 40 (1):226-244.

Kalimo, E. 2005. OECD Social Indicators for 2001: a critical appraisal. Social Indicators Research 70 (2):185-229.

Keeney, R. L., and R. S. Gregory. 2005. Selecting attributes to measure the achievement of objectives. Operations Research 53 (1):1-11.

Kildow, J. 2011. The Utility of Economic Indicators to Promote Policy-Relevant Science for Climate Change Decisions. In World Fisheries: A Social-Ecological Analysis, eds. R. E. Ommer, R. I. Perry, K. L. Cochrane and P. Cury, 137-150: Wiley-Blackwell.

Kingsley, G. T., and K. L. S. Pettit. 2011. Quality of Life at a Finer Grain: The National Neighborhood Indicators Partnership. In Community Quality-of-Life Indicators: Best Cases V, eds. M. J. Sirgy, R. Phillips and D. Rahtz. Dordrecht, Heidelberg, London, New York: Springer.

Kissinger, M., and W. E. Rees. 2010. An interregional ecological approach for modelling sustainability in a globalizing world--Reviewing existing approaches and emerging directions. Ecological Modelling 221 (21):2615-2623.

Kitzes, J., A. Galli, M. Bagliani, J. Barrett, G. Dige, S. Ede, K. Erb, S. Giljum, H. Haberl, C. Hails, L. Jolia-Ferrier, S. Jungwirth, M. Lenzen, K. Lewis, J. Loh, N. Marchettini, H. Messinger, K. Milne, R. Moles, C. Monfreda, D. Moran, K. Nakano, A. Pyhälä, W. Rees, C. Simmons, M. Wackernagel, Y. Wada, C. Walsh, and T. Wiedmann. 2009. A research agenda for improving national Ecological Footprint accounts. Ecological Economics 68 (7):1991-2007.

Kitzes, J., and M. Wackernagel. 2009. Answers to common questions in Ecological Footprint accounting. Ecological Indicators 9 (4):812-817.

Klevmarken, N. A. 2009. Towards an applicable true cost-of-living index that incorporates housing. Journal of Economic and Social Measurement 34 (1):19-33.

Kling, D., and J. N. Sanchirico. 2009. An Adaptation Portfolio for the United States Coastal and Marine Environment. In Adaptation: An Initiative of the Climate Policy Program at RFF: Resources for the Future.

Knight, K. W., and E. A. Rosa. 2011. The environmental efficiency of well-being: a cross-national analysis. Social Science Research 40 (3):931-949.

Kopytko, N., and J. Perkins. 2011. Climate change, nuclear power, and the adaptation–mitigation dilemma. Energy Policy 39:318-333.

Kuehn, B. M. 2009. CDC links data on health and environment. JAMA 302 (10):1049.

Kulig, A., H. Kolfoort, and R. Hoekstra. 2010. The case for the hybrid capital approach for the measurement of the welfare and sustainability. Ecological Indicators 10 (2):118-128.

Lake, R. W. 2000. Contradictions at the local scale: local implementation of Agenda 21 in the USA. In Consuming Cities: the Urban Environment in the Global Economy after the Rio Declaration, eds. N. Low, B. Gleeson, I. Elander and R. Lidskog, 71-91. New York: Routledge.

Land, K., V. Lamb, and H. Zheng. 2011. How are the kids doing? How do we know? Social Indicators Research 100 (3):463-477.

Land, K. C. 1971. On the definition of social indicators. The American Sociologist 6 (4):322-325.

———. 1975. The role of quality of employment indicators in general social reporting systems. American Behavioral Scientist 18 (3):304-332.

———. 2001. Models and indicators. Social Forces 80 (2):381-410.

Land, K. C., V. L. Lamb, S. O. Meadows, and A. Taylor. 2007. Measuring trends in child well-being: an evidence-based approach. Social Indicators Research 80 (1):105-132.

Lane, M. E., P. H. Kirshen, and R. M. Vogel. 1999. Indicators of impacts of global climate change on U.S. water resources. Journal of Water Resources Planning and Management 125 (4):194-204.

Lawless, N. M., and R. E. Lucas. 2011. Predictors of regional well-being: a county level analysis. Social Indicators Research 101 (3):341-357.

Lawn, P., and M. Clarke. 2006. Comparing Victoria's Genuine Progress with that of the Rest-of-Australia. Journal of Economic and Social Policy 10 (2):Article 7.

———. 2010. The end of economic growth? A contracting threshold hypothesis. Ecological Economics 69 (11):2213-2223.

Lawn, P. A. 2003. A theoretical foundation to support the Index of Sustainable Economic Welfare (ISEW), Genuine Progress Indicator (GPI), and other related indexes. Ecological Economics 44:105-118.

Lawrence, P., J. Meigh, and C. Sullivan. 2002. The Water Poverty Index: an international comparison. In Keele Economics Research Papers: Keele University.

Lee, J., V. Lamb, and K. Land. 2009. Composite indices of changes in child and youth well-being in the San Francisco Bay Area and the state of California, 1995–2005. Child Indicators Research 2 (4):353-373.

Lerner, M. 1973. Conceptualization of health and social well-being. Health Services Research 8 (1):6-12.

———. 1979. A review of "Health: United States, 1975". Social Indicators Research 6 (2):197-206.

Lerner-Lam, A. 2007. Assessing global exposure to natural hazards: progress and future trends. Environmental Hazards 7 (1):10-19.

Levine, J. N., A.-M. Esnard, and A. Sapat. 2007. Population displacement and housing dilemmas due to catastrophic disasters. Journal of Planning Literature 22 (1):3-15.

Lewis, K., and S. Burd-Sharps. 2010. A Century Apart: New Measures of Well-Being for U.S. Racial and Ethnic Groups. In American Human Development Project: SSRC.

Li, J., and B. Dawson. 2008. From Patchwork to National Network: Working Collaboratively to Create a National Environmental Public Health Tracking Network. Journal of Public Health Management & Practice 14 (6):596-599.

Liverman, D. M., and R. M. R. Cuesta. 2008. Human interactions with the Earth system: people and pixels revisited. Earth Surface Processes and Landforms 33 (9):1458-1471.

Lockwood, B., J. Martin, C. Y. Cao, and M. Schmitt. 2011. The Metropolitan Philadelphia Indicators Project: Measuring a Diverse Region. In Community Quality-of-Life Indicators: Best Cases V, eds. M. J. Sirgy, R. Phillips and D. Rahtz, 137-154. Dordrecht, Heidelberg, London, New York: Springer

Lyytimäki, J., and U. Rosenström. 2008. Skeletons out of the closet: effectiveness of conceptual frameworks for communicating sustainable development indicators. Sustainable Development 16 (5):301-313.

MacKerron, G. 2011. Happiness economics from 35 000 feet. Journal of Economic Surveys (in press).

Macrae Jr., D. 1989. Policy indicators: a continuing and needed field. Journal of Public Policy 9 (4):437-438.

Magis, K. 2010. Community resilience: an indicator of social sustainability. Society & Natural Resources: An International Journal 23 (5):401-416.

Malecki, K. C., B. Resnick, and T. A. Burke. 2008. Effective environmental public health surveillance programs: a framework for identifying and evaluating data resources and indicators. Journal of Public Health Management & Practice 14 (6):543-551.

Malghan, D. 2011. A dimensionally consistent aggregation framework for biophysical metrics. Ecological Economics 70 (5):900-909.

Malkina-Pykh, I. G., and Y. A. Pykh. 2008. Quality-of-life indicators at different scales: theoretical background. Ecological Indicators 8 (6):854-862.

Marks, E., M. D. Cargo, and M. Daniel. 2007. Constructing a health and social indicator framework for indigenous community health research. Social Indicators Research 82 (1):93-110.

Mascarenhas, A., P. Coelho, E. Subtil, and T. B. Ramos. 2010. The role of common local indicators in regional sustainability assessment. Ecological Indicators 10 (3):646-656.

Mayer, A. L. 2008. Strengths and weaknesses of common sustainability indices for multidimensional systems. Environment International 34 (2):277-291.

Mayer, A. L., C. W. Pawlowski, and H. Cabezas. 2006. Fisher Information and dynamic regime changes in ecological systems. Ecological Modelling 195 (1-2):72-82.

Mazur, L., C. Milanes, K. Randles, and D. Siegel. 2010. Indicators of Climate Change in California: Environmental Justice Impacts: Integrated Risk Assessment Branch, Office of Environmental Health Hazard Assessment, Government of California.

McBride, A. C., V. H. Dale, L. M. Baskaran, M. E. Downing, L. M. Eaton, R. A. Efroymson, C. T. Garten Jr., K. L. Kline, H. I. Jager, P. J. Mulholland, E. S. Parish, P. E. Schweizer, and J. M. Storey. 2011. Indicators to support environmental sustainability of bioenergy systems. Ecological Indicators 11 (5):1277-1289.
McCool, S. F., and G. H. Stankey. 2004. Indicators of sustainability: challenges and opportunities at the interface of science and policy. Environmental Management 33 (3):294-305.

McGeehin, M. A. 2008. National Environmental Public Health Tracking program: providing data for sound public health decisions. Journal of Public Health Management and Practice 14 (6):505-506.

McGeehin, M. A., J. R. Qualters, and A. S. Niskar. 2004. National Environmental Public Health Tracking Program: bridging the information gap. Environmental Health Perspectives 112 (14):1409-1413.

McGillivray, M., and A. Shorrocks. 2005. Inequality and multidimensional well-being. Review of Income and Wealth 51 (2):193-199.

McGinnis, R. 1979. Science indicators/1976: a critique. Social Indicators Research 6 (2):163-180.

Mclaughlin, S., and J. A. G. Cooper. 2010. A multi-scale coastal vulnerability index: A tool for coastal managers? Environmental Hazards 9 (3):233-248.

Meadows, D. 1998. Indicators and Information Systems for Sustainable Development. Hartland, VT: The Sustainability Institute.

Mendes, J. M. d. O. 2009. Social vulnerability indexes as planning tools: beyond the preparedness paradigm. Journal of Risk Research 12 (1):43-58.

Metropolitan Philadelphia Indicators Project (MPIP). 2010. Where we stand: Community indicators for Metropolitan Philadelphia. Philadelphia: Temple University.

Michael, A., D. Damani, D. DiFrancesco, J. Dickson, G. Seto, S. Tee, M. Sterniczuk, and D. Anastos. 2009. Indian and Northern Affairs Canada Thematic Indicators Project.

Michalos, A. C. 2010. What did Stiglitz, Sen and Fitoussi get right and what did they get wrong? Social Indicators Research 102 (1):117-129.

Millennium Challenge Corporation (MCC). 2011. Guide to the MCC Indicators and the Selection Process: Fiscal Year 2011: MCC.

Mills, D., and K. Ebi. 2011. Vulnerability White Paper: In Support of the National Climate Assessment's Vulnerability Assessment Workshop (Final Report), 49pp. Washington, DC: USGCRP National Climate Assessment.

Mills, W. B., C.-F. Chung, and K. Hancock. 2005. Predictions of Relative Sea-Level Change and Shoreline Erosion over the 21st Century on Tangier Island, Virginia. Journal of Coastal Research 21 (2):e36 – e51.

Mitchell, R. E., and J. R. Parkins. 2011. The challenge of developing social indicators for cumulative effects assessment and land use planning. Ecology and Society 16 (2):29 [online].

Moffatt, I. 1999. Is Scotland sustainable? A the series of indicators of sustainable development. International Journal of Sustainable Development and World Ecology 6 (4):242-250.

———. 2008. A preliminary analysis of composite indicators of sustainable development. International Journal of Sustainable Development and World Ecology 15 (2):81-87.

Moffatt, I., and M. Wilson. 1994. An Index of Sustainable Economic Welfare for Scotland, 1980-1991. International Journal of Sustainable Development and World Ecology 1 (4):264-291.

Molle, F., and P. Mollinga. 2003. Water poverty indicators: conceptual problems and policy issues. Water Policy 5 (5):529-544.

Montero, J.-M., C. Chasco, and B. Larraz. 2010. Building an environmental quality index for a big city: a spatial interpolation approach combined with a distance indicator. Journal of Geographical Systems 12 (4):435-459.

Moore, D., G. Cranston, A. Reed, and A. Galli. 2011. Projecting future human demand on the Earth's regenerative capacity. Ecological Indicators ((in press)).

Moran, D. D., M. Wackernagel, J. A. Kitzes, S. H. Goldfinger, and A. Boutaud. 2008. Measuring sustainable development — Nation by nation. Ecological Economics 64 (3):470-474.

Morin, A. 2005. The Canadian Water Sustainability Index (CWSI) (PRI Working Paper Series #011). Ottawa: Policy Research Initiative.

Morse, S. 2006. Is corruption bad for environmental sustainability? a cross-national analysis. Ecology and Society 11 (1):22.

———. 2008. On the use of headline indices to link environmental quality and income at the level of the nation state. Applied Geography 28 (2):77-95.

———. 2008. The geography of tyranny and despair: development indicators and the hypothesis of genetic inevitability of national inequality. Geographical Journal 174 (3):195-206.

———. 2010. Out of sight, out of mind. Reporting of three indices in the UK national press between 1990 and 2009. Sustainable Development (published online).

Morse, S., and E. D. G. Fraser. 2005. Making 'dirty' nations look clean? The nation state and the problem of selecting and weighting indices as tools for measuring progress towards sustainability. Geoforum 36 (5):625-640.

Morse, S., N. McNamara, M. Acholo, and B. Okwoli. 2010. Sustainability indicators: the problem of integration. Sustainable Development 9 (1):1-15.

Moser, A. 2007. Gender and Indicators: Overview Report. In BRIDGE Development -- Gender: UNDP.

Moser, S., G. Franco, S. Pittiglio, W. Chou, and D. Cayan. 2009. The future is now: an update on climate change science impacts and response options for California: California Energy Commission, PIER Energy-Related Environmental Research Program. CEC-500-2008-071.

Moser, S. C. 2010. Now more than ever: the need for more societally relevant research on vulnerability and adaptation to climate change. Applied Geography 30 (4):464-474.

Moser, S. C., and A. L. Luers. 2008. Managing climate risks in California: the need to engage resource managers for successful adaptation to change. Climatic Change 87 (1):309-322.

Mosquera-Machado, S., and M. Dilley. 2009. A comparison of selected global disaster risk assessment results. Natural Hazards 48 (3):439-456.

Moss, R. H. 2007. Improving information for managing an uncertain future climate. Global Environmental Change 17:4-7.

Munda, G., M. Nardo, M. Saisana, and T. Srebotnjak. 2009. Measuring uncertainties in composite indicators of sustainability. International Journal of Environmental Technology and Management 11 (1-3):7-26.

Munier, N. 2011. Methodology to select a set of urban sustainability indicators to measure the state of the city, and performance assessment. Ecological Indicators 11 (5):1020-1026.

Muñoz-Erickson, T. A., B. Aguilar-González, and T. D. Sisk. 2007. Linking ecosystem health indicators and collaborative management: a systematic framework to evaluate ecological and social outcomes. Ecology and Society 12 (2):6 (online).

National Research Council (NRC). 1999. The Impacts of Natural Disasters: A Framework for Loss Estimation. Washington, D.C.: National Academy Press.

———. 1999. Our Common Journey: A Transition Toward Sustainability. Washington, D.C.: National Academy Press.

———. 2006. Drawing Louisiana's New Map: Addressing Land Loss in Coastal Louisiana. Washington, D.C.: National Academies Press.

———. 2010. Monitoring Climate Change Impacts: Metrics at the Intersection of the Human and Earth Systems. Washington, D.C.: National Academies Press.

———. 2010. America's Climate Choices: Adapting to the Impacts of Climate Change. Washington, D.C.: National Academies Press.

Natoli, R. 2008. Indicators of economic and social progress: an assessment and an alternative, School of Applied Economics, Faculty of Business and Law, Victoria University, Melbourne, Australia.

Natoli, R., and S. Zuhair. 2009. Human Progress: Concepts and Measurements: VDM Verlag.

———. 2010. What is a reasonable measure of progress? The International Journal of Sociology and Social Policy 30 (5/6):201-218.

———. 2010. Establishing the RIE index: a review of the components critical to progress measurement. International Journal of Social Economics 37 (8):574-591.

———. 2011. Measuring progress: a comparison of the GDP, HDI, GS and the RIE. Social Indicators Research 103 (1):33-56.

Ness, B., S. Anderberg, and L. Olsson. 2010. Structuring problems in sustainability science: The multi-level DPSIR framework. Geoforum 41 (3):479-488.

Ness, B., E. Urbel-Piirsalu, S. Anderberg, and L. Olsson. 2007. Categorising tools for sustainability assessment. Ecological Economics 60 (3):498-508.

Neumayer, E. 2000. On the methodology of ISEW, GPI and related measures: some constructive suggestions and some doubt on the 'threshold' hypothesis. Ecological Economics 34 (3):347-361.

Neumayer, E., and F. Barthel. 2011. Normalizing economic loss from natural disasters: A global analysis. Global Environmental Change 21:13-24.

New, M., D. Liverman, H. Schroder, and K. Anderson. 2011. Four degrees and beyond: the potential for a global temperature increase of four degrees and its implications. Philosophical Transactions of the Royal Society A 369 (1934):6-19.

Ng, Y.-K. 2008. Environmentally responsible Happy Nation Index: towards an internationally acceptable national success indicator. Social Indicators Research 85 (3):425-446.

————. 2008. Happiness Studies: Ways to Improve Comparability and Some Public Policy Implications. Economic Record 84 (265):253-266.

Nicholls, R., P. Wong, V. Burkett, C. Woodroffe, and J. Hay. 2008. Climate change and coastal vulnerability assessment: scenarios for integrated assessment. Sustainability Science 3 (1):89-102.

Niemeijer, D. 2002. Developing indicators for environmental policy: data-driven and theory-driven approaches examined by example. Environmental Science & Policy 5 (2):91-103.

Niemeijer, D., and R. S. d. Groot. 2008. A conceptual framework for selecting environmental indicator sets. Ecological Indicators 8 (1):14-25.

————. 2008. Framing environmental indicators: moving from causal chains to causal networks. Environment, Development and Sustainability 10 (1):89-106.

Nkonya, E., M. Winslow, M. S. Reed, M. Mortimore, and A. Mirzabaev. 2011. Monitoring and assessing the influence of social, economic and policy factors on sustainable land management in drylands. Land Degradation & Development 22 (2):240-247.

Noll, H.-H. 2011. The Stiglitz-Sen-Fitoussi-Report: Old wine in new skins? Views from a social indicators perspective. Social Indicators Research 102 (1):111-116.

Norman, E. S., K. Bakker, and G. Dunn. 2011. Recent Developments in Canadian Water Policy: An Emerging Water Security Paradigm. Canadian Water Resources Journal 36 (1):53-66.

Nourry, M. 2008. Measuring sustainable development: Some empirical evidence for France from eight alternative indicators. Ecological Economics 67 (3):441-456.

O'Neill, M. S., D. K. Jackman, M. Wyman, X. Manarolla, C. J. Gronlund, D. G. Brown, S. J. Brines, J. Schwartz, and A. V. Diez-Roux. 2010. US local action on heat and health: are we prepared for climate change? International Journal of Public Health 55 (2):105-112.

OECD. 2003. OECD Environmental Indicators: Development, Measurement and Use. Paris: Organisation for Economic Co-operation and Development.

————. 2011. Society at a Glance 2011- OECD Social Indicators (www.oecd.org/els/social/indicators/ SAG).

Officer, L. H. 2007. An Improved Long-Run Consumer Price Index for the United States. Historical Methods: A Journal of Quantitative and Interdisciplinary History 40 (3):135-148.

————. 2010. The Annual Consumer Price Index for the United States, 1774-2010. In MeasuringWorth.

Ollivier, T., and P.-N. Giraud. 2010. The Usefulness of Sustainability Indicators for Policy Making in Developing Countries: The Case of Madagascar. Journal of Environment & Development 19 (4):399-423.

Osberg, L. 2009. Measuring economic security in insecure times: new perspectives, new events, and the Index of Economic Well-Being: Centre for the Study of Living Standards.

Osberg, L., and A. Sharpe. 2002. An Index of Economic Well-Being for Selected OECD countries. Review of Income and Wealth 48 (3):291-316.

————. 2009. New Estimates of the Index of Economic Well-being for Selected OECD Countries, 1980-2007: Centre for the Study of Living Standards.

Oswald, A. J., and S. Wu. 2010. Objective Confirmation of Subjective Measures of Human Well-Being: Evidence from the U.S.A. Science 327 (5965):576-579.

Oxfam America. 2009. Exposed: Social Vulnerability and Climate Change in the US Southeast. Boston, MA: Oxfam America Inc.

Özdemir, E. D., M. Härdtlein, T. Jenssen, D. Zech, and L. Eltrop. 2011. A confusion of tongues or the art of aggregating indicators—Reflections on four

projective methodologies on sustainability measurement. Renewable and Sustainable Energy Reviews 15:2385-2396.

Özyurt, G., and A. Ergin. 2010. Improving coastal vulnerability assessments to sea-level rise: a new indicator-based methodology for decision makers. Journal of Coastal Research 26 (2):265-273.

Pais, J. F., and J. R. Elliot. 2008. Places as recovery machines: vulnerability and neighborhood change after major hurricanes. Social Forces 86 (4):1415-1453.

Palumbi, S. R., P. A. Sandifer, J. D. Allan, M. W. Beck, D. G. Fautin, M. J. Fogarty, B. S. Halpern, L. S. Incze, J.-A. Leong, E. Norse, J. J. Stachowicz, and D. H. Wall. 2009. Managing for ocean biodiversity to sustain marine ecosystem services. . Frontiers in Ecology and the Environment 7 (4):204–211.

Parke, R., and D. Seidman. 1978. Social indicators and social reporting. The Annals of the American Academy of Political and Social Science 435 (1):1-16.

Parris, T. M., and R. W. Kates. 2003. Characterizing and measuring sustainable development. Annual Review of Environment and Resources 28:559-586.

Patridge, J., and G. Namulanda. 2008. Describing environmental public health data: implementing a descriptive metadata standard on the Environmental Public Health Tracking Network. Journal of Public Health Management & Practice 14 (6):515-525.

Patt, A. G., M. Tadross, P. Nussbaumer, K. Asante, M. Metzger, J. Rafael, A. Goujon, and G. Brundrit. 2010. Estimating least-developed countries' vulnerability to climate-related extreme events over the next 50 years. Proceedings of the National Academy of Sciences 107 (4):1333-1337.

Peduzzi, P., H. Dao, C. Herold, and F. Mouton. 2009. Assessing global exposure and vulnerability towards natural hazards: the Disaster Risk Index. Natural Hazards and Earth System Sciences 9:1149-1159.

Pendleton, E. A., E. R. Thieler, and S. J. Williams. 2010. Importance of Coastal Change Variables in Determining Vulnerability to Sea- and Lake-Level Change. Journal of Coastal Research 26 (1):176-183.

Perch-Nielsen, S. L. 2010. The vulnerability of beach tourism to climate change—an index approach. Climatic Change 100 (3-4):579-606.

Phillis, Y. A., E. Grigoroudis, and V. S. Kouikoglou. 2011. Sustainability ranking and improvement of countries. Ecological Economics 70 (3):542-553.

Pillarisetti, J. R., and J. C. J. M. van den Bergh. 2010. Sustainable nations: what do aggregate indexes tell us? Environment, Development and Sustainability 12 (1):49-62.

Pine, J. C. 2008. Natural Hazards Analysis: Reducing the Impact of Disasters. Boca Raton, FL: Auerbach Publications, Taylor & Francis Group.

Pintér, L., P. Hardi, and P. Bartelmus. 2005. Sustainable Development Indicators: Proposals for a way forward: IISD commissioned by the United Nations Division for Sustainable Development (UN-DSD).

PlaNYC. 2011. PlaNYC Update April 2011. New York, NY: The City of New York.

Policy Research Initiative (PRI). 2007. Canadian Water Sustainability Index. Sustainable Development Briefing Note. Ottawa: Policy Research Initiative, Government of Canada.

————. 2007. Canadian Water Sustainability Index (CWSI) Project Report. Ottawa: Policy Research Initiative, Government of Canada.

Polsky, C., R. Neff, and B. Yarnal. 2007. Building comparable global change vulnerability assessments: The vulnerability scoping diagram. Global Environmental Change 17 (3-4):472-485.

Ponce-Hernandez, R., and P. Koohafkan. 2010. A Methodology for Land Degradation Assessment at Multiple Scales Based on the DPSIR Approach: Experiences from Applications to Drylands. In Land Degradation and Desertification: Assessment, Mitigation and Remediation, ed. P. Z. e. al., 49-65: Springer.

Ponthiere, G. 2009. The ecological footprint: an exhibit at an intergenerational trial? Environment, Development and Sustainability 11 (4):677-694.

Prescott-Allen, R. 2001. The Well-Being of Nations: A country-by-country index of quality of life and the environment. Washington, D.C.: Island Press.

Pülzl, H., and E. Rametsteiner. 2009. Indicator development as 'boundary spanning' between scientists and policy-makers. Science and Public Policy 36 (10):743-752.

Quay, R., and K. Hutanuwatr. 2009. Visualization of Sustainability Indicators: A Conceptual Framework. In Visualizing Sustainable Planning, eds. H. Hagen, S. Guhathakurta and G. Steinebach, 203-213. Berlin and Heidelberg: Springer.

Rae, A. 2009. Isolated entities or integrated neighbourhoods? An alternative view of the measurement of deprivation. Urban Studies 46 (9):1859-1878.

Rahman, T., R. C. Mittelhammer, and P. Wandschneider. 2005. Measuring the Quality of Life across countries: a sensitivity analysis of well-being indices. In UNU-WIDER conference on Inequality, Poverty and Human Well-being. Helsinki, Findland: United Nations University World Institute for Development Economics Research.

Rametsteiner, E., H. Pülzl, J. Alkan-Olsson, and P. Frederiksen. 2011. Sustainability indicator development--Science or political negotiation? Ecological Indicators 11 (1):61-70.

Ramos, T. B. 2009. Development of regional sustainability indicators and the role of academia in this process: the Portuguese practice. Journal of Cleaner Production 17:1101-1115.

Ramos, T. B., and S. Caeiro. 2010. Meta-performance evaluation of sustainability indicators. Ecological Indicators 10 (2):157-166.

Raskin, P. D., C. Electris, and R. A. Rosen. 2010. The century ahead: searching for sustainability. Sustainability 2 (8):2626-2651.

Ravallion, M. 2010. Mashup indices of development. In Policy Research Working Paper 5432: The World Bank Development Research Group.

Ray, A. K. 2008. Measurement of social development: an international comparison. Social Indicators Research 86 (1):1-46.

Reed, M., E. D. G. Fraser, S. Morse, and A. J. Dougill. 2005. Integrating methods for developing sustainability indicators to facilitate learning and action. Ecology and Society 10 (1):r3 [online] URL: http://www.ecologyandsociety.org/vol10/iss1/resp3/.

Reed, M. S., E. D. G. Fraser, and A. J. Dougill. 2006. An adaptive learning process for developing and applying sustainability indicators with local communities. Ecological Economics 59 (4):406-418.

Renn, D., N. Pfaffenberger, M. Platter, H. Mitmansgruber, R. Cummins, and S. Höfer. 2009. International Well-being Index: The Austrian Version. Social Indicators Research 90 (2):243-256.

Reynolds, J. F., A. Grainger, D. M. Stafford Smith, G. Bastin, L. Garcia-Barrios, R. J. Fernández, M. A. Janssen, N. Jürgens, R. J. Scholes, A. Veldkamp, M. M. Verstraete, G. Von Maltitz, and P. Zdruli. 2011. Scientific concepts for an integrated analysis of desertification. Land Degradation & Development 22 (2):166-183.

Reynolds, J. F., D. M. S. Smith, E. F. Lambin, B. L. T. II, M. Mortimore, S. P. J. Batterbury, T. E. Downing, H. Dowlatabadi, R. J. Fernández, J. E. Herrick, E. Huber-Sannwald, H. Jiang, R. Leemans, T. Lynam, F. T. Maestre, M. Ayarza, and B. Walker. 2007. Global desertification: building a science for dryland development. Science 316:847-851.

Ritz, B., I. Tager, and J. Balmes. 2005. Can lessons from public health disease surveillance be applied to environmental public health tracking? Environmental Health Perspectives 113 (3):243-249.

Roberts, N. J., F. Nadim, and B. Kalsnes. 2009. Quantification of vulnerability to natural hazards. Georisk: Assessment and Management of Risk for Engineered Systems and Geohazards 3 (3):164-173.

Romieu, E., T. Welle, S. Schneiderbauer, M. Pelling, and C. Vinchon. 2010. Vulnerability assessment within climate change and natural hazard contexts: revealing gaps and synergies through coastal applications. Sustainability Science 5 (2):159-170.

Roth, E., H. Rosenthal, and P. Burbridge. 2000. A discussion of the use of the sustainability index: 'ecological footprint' for aquaculture production. Aquatic Living Resources 13:461-469.

Rygel, L., D. O'Sullivan, and B. Yarnal. 2006. A method for constructing a Social Vulnerability Index: an application to hurricane storm surges in a developed country. Mitigation and Adaptation Strategies for Global Change 11 (3):741-764.

Saisana, M., and S. Tarantola. 2002. State-of-the-art report on current methodologies and practices for composite indicator development. Ispra, Italy: Institute for the Protection and Security of the Citizen, Technological and Economic Risk Management.

Saltelli, A. 2007. Composite Indicators between Analysis and Advocacy. Social Indicators Research 81 (1):65-77.

Salvaris, M. 2000. Community and social indicators: How citizens can measure progress. Hawthorn, Australia: Institute for Social Research.

Salvati, L., and M. Zitti. 2009. Substitutability and weighting of ecological and economic indicators: exploring the importance of various components of a synthetic index. Ecological Economics 68 (4):1093-1099.

Sawicki, D. S. 2002. Improving Community Indicator Systems: Injecting More Social Science into the Folk Movement. Planning Theory & Practice 3 (1):13-32.

Sawicki, D. S., and P. Flynn. 1996. Neighborhood indicators: a review of the literature and an assessment of conceptual and methodological issues. Journal of the American Planning Association 62 (2):165-183.

Schelfaut, K., B. Pannemans, I. van der Craats, J. Krywkow, J. Mysiak, and J. Cools. 2011. Bringing flood resilience into practice: the FREEMAN project. Environmental Science & Policy (In press).

Schepelmann, P., Y. Goossens, and A. Makipaa eds. 2010. Towards sustainable development: alternatives to GDP for measuring progress: Wuppertal Institute for Climate, Environment and Energy.

Schlossberg, M., and A. Zimmerman. 2003. Developing statewide indices of environmental, economic, and social sustainability: a look at Oregon and the Oregon Benchmarks. Local Environment: The International Journal of Justice and Sustainability 8 (6):641-660.

Schmidtlein, M. C. 2008. Spatio-temporal changes in the social vulnerability of Charleston, South Carolina from 1960 to 2010, Dissertation, Department of Geography, University of South Carolina.

Schmidtlein, M. C., R. C. Deutsch, W. W. Piegorsch, and S. L. Cutter. 2008. A sensitivity analysis of the Social Vulnerability Index. Risk Analysis 28 (4):1099-1114.

Schmidtlein, M. C., C. Finch, and S. L. Cutter. 2008. Disaster declarations and major hazard occurrences in the United States. The Professional Geographer 60 (1):1-14.

Schröter, D., C. Polsky, and A. G. Patt. 2005. Assessing vulnerabilities to the effects of global change: an eight step approach. Mitigation and Adaptation Strategies for Global Change 10 (4):573-595.

Scipioni, A., A. Mazzi, M. Mason, and A. Manzardo. 2009. The Dashboard of Sustainability to measure the local urban sustainable development: The case study of Padua Municipality. Ecological Indicators 9 (2):364-380.

Segnestam, L. 2002. Indicators of environment and sustainable development: theories and practical experience. In Environmental Economics Series, Paper No. 89: World Bank Environment Department.

Semenza, J. C., G. B. Ploubidis, and L. A. George. 2011. Climate change and climate variability: personal motivation for adaptation and mitigation. Environmental Health 10:46.

Sempier, T. T., D. L. Swann, R. Emmer, S. H. Sempier, and M. Schneider. 2010. Coastal Resilience Index: A Community Self-Assessment. MASGP-08-014: National Oceanic and Atmospheric Administration, Mississippi-Alabama Sea Grant Consortium, and the Gulf of Mexico Alliance Coastal Community Resilience Team.

Sharpe, A. 1999. A survey of indicators of economic and social well-being. In Paper prepared for Canadian Policy Research Networks: Centre for the Study of Living Standards.

Sharpe, A., A. Ghanghro, E. Johnson, and A. Kidwai. 2011. Does money matter? Determining the happiness of Canadians: Centre for the Study of Living Standards.

Sheffield, P. E., and P. J. Landrigan. 2011. Global climate change and children's health: threats and strategies for prevention. Environmental Health Perspectives 119 (3):291-298.

Shen, L.-Y., J. Jorge Ochoa, M. N. Shah, and X. Zhang. 2011. The application of urban sustainability indicators - A comparison between various practices. Habitat International 35 (1):17-29.

Sherrieb, K., F. Norris, and S. Galea. 2010. Measuring capacities for community resilience. Social Indicators Research 99 (2):227-247.

Shields, D. J., S. V. Solar, and W. E. Martin. 2002. The role of values and objectives in communicating indicators of sustainability. Ecological Indicators 2 (1-2):149-160.

Shire, J. D., G. M. Marsh, E. O. Talbott, and R. S. Sharma. 2011. Advances and current themes in occupational health and environmental public health surveillance. Annual Review of Public Health 32:109-132.

Siche, J. R., F. Agostinho, E. Ortega, and A. Romeiro. 2008. Sustainability of nations by indices: Comparative study between environmental sustainability index, ecological footprint and the emergy performance indices. Ecological Economics 66 (4):628-637.

Simpson, D. M., and M. Katirai. 2006. Indicator issues and proposed framework for a Disaster Preparedness Index (DPi). Working Paper 06-03. Louisville, KY: Center for Hazards Research and Policy Development, School of Urban and Public Affairs, University of Louisville.

Singh, R. K., H. R. Murty, S. K. Gupta, and A. K. Dikshit. 2009. An overview of sustainability assessment methodologies. Ecological Indicators 9 (2):189-212.

Sirgy, M. J., A. C. Michalos, A. L. Ferriss, R. A. Easterlin, D. Patrick, and W. Pavot. 2006. The Quality-of-Life (QOL) research movement: past, present, and future. Social Indicators Research 76 (3):343-466.

Skaaning, S.-E. 2010. Measuring the rule of law. Political Research Quarterly 63 (2):449-460.

Smit, B., G. K. Hovelsrud, J. Wandel, and M. Andrachuk. 2010. Introduction to the CAVIAR Project and Framework. In Community Adaptation and Vulnerability in Arctic Regions, eds. G. K. Hovelsrud and B. Smit, 1-22: Springer Netherlands.

Smit, B., and J. Wandel. 2006. Adaptation, adaptive capacity and vulnerability. Global Environmental Change 16 (3):282-292.

Smithers, J., and B. Smit. 1997. Human adaptation to climatic variability and change. Global Environmental Change 7 (2):129-146.

Soares Jr., J., and R. H. Quintella. 2008. Development: an analysis of concepts, measurement and indicators. Brazilian Adminstration Review 5 (2):104-124.

Solecki, W., R. Leichenko, and K. O'Brien. 2011. Climate change adaptation strategies and disaster risk reduction in cities: connections, contentions, and synergies. Current Opinion in Environmental Sustainability 3 (3):135-141.

Somarriba, N., and B. Pena. 2009. Synthetic indicators of quality of life in Europe. Social Indicators Research 94 (1):115-133.

Sommer, S., C. Zucca, A. Grainger, M. Cherlet, R. Zougmore, Y. Sokona, J. Hill, R. Della Peruta, J. Roehrig, and G. Wang. 2011. Application of indicator systems for monitoring and assessment of desertification from national to global scales. Land Degradation & Development 22 (2):184-197.

Sonntag, V. 2010. Designing sustainability indicator frameworks for information flow: a case study of B-Sustainable. Applied Research in Quality of Life 5 (4):325-339.

South Pacific Applied Geoscience Commission (SOPAC), and United Nations Environment Programme (UNEP). 2004. EVI: Description of Indicators: SOPAC & UNEP.
————. 2005. Building Resilience in SIDS: The Environmental Vulnerability Index: SOPAC & UNEP.

Spangenberg, J. H. 2002. Institutional sustainability indicators: an analysis of the institutions in Agenda 21 and a draft set of indicators for monitoring their effectivity. Sustainable Development 10 (2):103-115.

Spangenberg, J. H., S. Pfahl, and K. Deller. 2002. Towards indicators for institutional sustainability: lessons from an analysis of Agenda 21. Ecological Indicators 2 (1-2):61-77.

Srebotnjak, T., G. Carr, A. de Sherbinin, and C. Rickwood. 2011. A global Water Quality Index and hot-deck imputation of missing data. Ecological Indicators (in press).

Staniūnas, M., E. K. Staniūnas, and M. Burinskiené. 2010. Application of indices for assessing the ecological potential of urban development. Ekologija 56 (3-4):79-86.

Steurer, R., and M. Hametner. 2011. Objectives and indicators in sustainable development strategies: similarities and variances across Europe. Sustainable Development 19:(In press).

Stiglitz, J. E., A. Sen, and J.-P. Fitoussi. 2009. Report by the Commission on the Measurement of Economic Performance and Social Progress. Paris.

Strzepek, K., G. Yohe, J. Neumann, and B. Boehlert. 2010. Characterizing changes in drought risk for the United States from climate change. Environmental Research Letters 5 (4):044012.

Stuby, R. G. 1979. Some new directions for social indicators in the U.S. Department of Agriculture. Social Indicators Research 6 (2):273-282.

Stutz, J. 2011. The Quality of Development Index – A New Headline Indicator of Progress. Journal of Future Studies 15 (3):73-102.

Subramanian, S. 2004. Indicators of inequality and poverty. Research Paper No. 2004/25. Helsinki, Finland: World Institute for Development Economics Research (WIDER), United Nations University (UNU).

Sullivan, C. 2002. Calculating a Water Poverty Index. World Development 30 (7):1195-1210.

Sullivan, C., and J. Meigh. 2005. Targeting attention on local vulnerabilities using an integrated index approach: the example of the climate vulnerability index. Water Science & Technology 51 (5):69-78.

Sullivan, C., J. Meigh, and P. Lawrence. 2006. Application of the Water Poverty Index at different scales: a cautionary tale. Water International 31 (3):412-426.

Sullivan, C. A. 2011. Quantifying water vulnerability: a multi-dimensional approach. Stochatic Environmental Research and Risk Assessment 25 (4):627-640.

Sullivan, C. A., J. R. Meigh, and A. M. Giacomello. 2003. The Water Poverty Index: Development and application at the community scale. Natural Resources Forum 27 (3):189-199.

Sutton, P. C., S. J. Anderson, B. T. Tuttle, and L. Morse. 2011. The real wealth of nations: mapping and monetizing the human ecological footprint. Ecological Indicators (In press).

Sydneysmith, R., M. Andrachuk, B. Smit, and G. K. Hovelsrud. 2010. Vulnerability and Adaptive Capacity in Arctic Communities. In Adaptive Capacity and Environmental Governance, eds. D. Armitage and R. Plummer, 133-156. Berlin and Heidelberg: Springer.

Taeuber, K. E. 1970. Toward a social report: a review article: Rand.

Talberth, J., C. Cobb, and N. Slattery. 2007. The Genuine Progress Indicator 2006: A Tool for Sustainable Development. Oakland, CA: Redefining Progress.

Tanguay, G. A., J. Rajaonson, J.-F. Lefebvre, and P. Lanoie. 2010. Measuring the sustainability of cities: An analysis of the use of local indicators. Ecological Indicators 10 (2):407-418.

Tapsell, S., S. McCarthy, H. Faulkner, and M. Alexander. 2010. Social vulnerability to natural hazards. CapHaz-Net WP4 Report. London, U.K.: Flood Hazard Research Centre, Middlesex University.

Tasaki, T., Y. Kameyama, S. Hashimoto, Y. Moriguchi, and H. Harasawa. 2010. A survey of national sustainable development indicators. International Journal of Sustainable Development 13 (4):337-361.

Tate, E., S. L. Cutter, and M. Berry. 2010. Integrated multihazard mapping. Environment and Planning B: Planning and Design 37 (4):646-663.

Tol, R. S. J., and G. W. Yohe. 2007. The weakest link hypothesis for adaptive capacity: an empirical test. Global Environmental Change 17 (2):218-227.

Tolbert, C. J., K. Mossberger, and R. McNeal. 2008. Institutions, Policy Innovation, and E-Government in the American States. Public Administration Review 68 (3):549-563.

Tsui, K.-Y. 2002. Multidimensional poverty indices. Social Choice and Welfare 19:69-93.

U.S. Department of Health Education and Welfare. 1969. Toward a Social Report. Washington, D.C.: U.S. Department of Health, Education and Welfare.

UNDP. 2009. Linking climate change policies to human development analysis and advocacy: a guidance note for Human Development Report teams: United Nations Development Programme.

UNEP. 2006. Environmental Indicators for North America: United Nations Environment Programme.

van de Kerk, G., and A. Manuel. 2010. Sustainable Society Index 2010: Sustainable Society Foundation.

van de Kerk, G., and A. R. Manuel. 2008. A comprehensive index for a sustainable society: the SSI — the Sustainable Society Index Ecological Economics 66 (2-3):228-242.

van den Bergh, J., and F. Grazi. 2010. On the policy relevance of ecological footprints. Environmental Science & Technology 44 (13):4843-4844.

van den Bergh, J., and H. Verbruggen. 1999. Spatial sustainability, trade and indicators: an evaluation of the 'ecological footprint'. Ecological Economics 29 (1):61-72.

van Koppen, C. S. A., A. P. J. Mol, and J. P. M. van Tatenhove. 2010. Coping with extreme climate events: institutional flocking. Futures 42 (7):749-758.

van Vuuren, D. P., M. Isaac, Z. W. Kundzewicz, N. Arnell, T. Barker, P. Criqui, F. Berkhout, H. Hilderink, J. Hinkel, A. Hof, A. Kitous, T. Kram, R. Mechler, and S. Scrieciu. 2011. The use of scenarios as the basis for combined assessment of climate change mitigation and adaptation. Global Environmental Change 21 (2):575-591.

Vancouver Foundation. 2010. Vital Signs for Metro Vancouver: 2010 On the road to vitality. Vancouver: Vancouver Foundation.

Vandivere, S., and C. McPhee. 2008. Methods for tabulating indices of child well-being and context: an illustration and comparison of performance in 13 American states. Child Indicators Research 1 (3):251-290.

Vemuri, A. W., and R. Costanza. 2006. The role of human, social, built, and natural capital in explaining life satisfaction at the country level: Toward a National Well-Being Index (NWI). Ecological Economics 58 (1):119-133.

Venetoulis, J., and C. Cobb. 2004. The Genuine Progress Indicator 1950-2002 (2004 Update). Oakland, CA: Redefining Progress.

Villa, F., and H. McLeod. 2002. Environmental vulnerability indicators for environmental planning and decision-making: guidelines and applications. Environmental Management 29 (3):335-348.

Vogel, C., S. C. Moser, R. E. Kasperson, and G. D. Dabelko. 2007. Linking vulnerability, adaptation, and resilience science to practice: pathways, players, and partnerships. Global Environmental Change 17 (3-4):349-364.

Vujakovic, P. 2010. How to Measure Globalization? A New Globalization Index (NGI). Atlantic Economic Journal 38 (2):237.

Wackernagel, M. 2009. Methodological advancements in footprint analysis. Ecological Economics 68 (7):1925-1927.

————. 2011. Global Footprint Network: Our Role in Ending Overshoot. In Earth Capitalism: Creating a New Civilization through a Responsible Market Economy, ed. P. U. Petit, 85-90. Piscataway, NJ: Transaction Publishers.

Wackernagel, M., C. Monfreda, N. B. Schulz, K.-H. Erb, H. Haberl, and F. Krausmann. 2004. Calculating national and global ecological footprint time series: resolving conceptual challenges. Land Use Policy 21 (3):271-278.

Wackernagel, M., and W. Rees. 1996. Our Ecological Footprint: Reducing Human Impact on the Earth. Philadelphia, PA, USA and Gabriola Island, BC, Canada: New Society Publishers.

Waddell, J., and M. Hambrick. 2010. Mapping social vulnerability to climate change in the U.S. Southeast. In Annual Meeting of the American Sociological Association. Hilton Atlanta and Atlanta Marriott Marquis, Atlanta, GA.

Ward, S. M., M. Leitner, and J. Pine. 2010. Investigating Recovery Patterns in Post Disaster Urban Settings: Utilizing Geospatial Technology to Understand Post-Hurricane Katrina Recovery in New Orleans, Louisiana. In Geospatial Techniques in Urban Hazard and Disaster Analysis, eds. P. S. Showalter and Y. Lu, 355-372: Springer Netherlands.

Weiland, U., A. Kindler, E. Banzhaf, A. Ebert, and S. Reyes-Paecke. 2011. Indicators for sustainable land use management in Santiago de Chile. Ecological Indicators 11 (5):1074-1083.

Weitzman, M. S. 1979. The developing program on social indicators at the U.S. Bureau of the Census. Social Indicators Research 6 (2):239-249.

Welfens, P., J. Perret, and D. Erdem. 2010. Global economic sustainability indicator: analysis and policy options for the Copenhagen process. International Economics and Economic Policy 7 (2):153-185.

Wells, A. K. 2006. The Boston Indicators Project : the role of indicators in supporting environmental efforts in the Boston metropolitan region, Dept. of Urban Studies and Planning, Massachusetts Institute of Technology.

Welsch, H. 2005. Constructing Meaningful Sustainability Indices. In Applied Research in Environmental Economics, eds. C. Böhringer and A. Lange, 7-22: Physica-Verlag HD.

Wen, Z., K. Zhang, B. Du, Y. Li, and W. Li. 2007. Case study on the use of genuine progress indicator to measure urban economic welfare in China. Ecological Economics 63 (2-3):463-475.

West, J. M., S. H. Julius, P. Kareiva, C. Enquist, J. J. Lawler, B. Petersen, A. E. Johnson, and M. R. Shaw. 2009. U.S. Natural Resources and Climate Change: Concepts and Approaches for Management Adaptation. Environmental Management 44 (6):1001-1021.

Wheeler, D. 2011. Quantifying Vulnerability to Climate Change: Implications for Adaptation Assistance: Center for Global Development Working Paper 240.

White, H. D. 1983. A cocitation map of the social indicators movement. Journal of the American Society for Information Science 34 (5):307-312.

Wiedmann, T., and J. Barrett. 2010. Review of the Ecological Footprint Indicator—Perceptions and Methods. Sustainability 2 (6):1645-1693.

Wiedmann, T., and M. Lenzen. 2007. On the conversion between local and global hectares in Ecological Footprint analysis. Ecological Economics 60:673-677.

Wiegand, J., D. Raffaelli, J. C. R. Smart, and P. C. L. White. 2010. Assessment of temporal trends in ecosystem health using an holistic indicator. Journal of Environmental Management 91 (7):1446-1455.

Wilder, M., C. A. Scott, N. P. Pablos, R. G. Varady, G. M. Garfin, and J. McEvoy. 2010. Adapting across boundaries: climate change, social learning, and resilience in the U.S.-Mexico border region. Annals of the Association of American Geographers 100 (4):917-928.

Wilhite, D. A., M. J. Hayes, C. Knutson, and K. H. Smith. 2000. Planning for drought: moving from crisis to risk management. Journal of the American Water Resources Association 36 (4):697-710.

Wilhite, D. A., M. D. Svoboda, and M. J. Hayes. 2007. Understanding the complex impacts of drought: a key to enhancing drought mitigation and preparedness. Water Resources Management 21 (5):763-774.

Wilson, J., P. Tyedmers, and R. Pelot. 2007. Contrasting and comparing sustainable development indicator metrics. Ecological Indicators 7 (2):299-314.

Wongbusarakum, S., and C. Loper. 2011. Indicators to assess community-level social vulnerability to climate change: An addendum to SocMon and SEM-Pasifika regional socioeconomic monitoring guidelines (First Draft for Public Circulation and Field Testing): CRISP, IUCN, TNC, SocMon, NOAA.

Yohe, G. W., and R. S. J. Tol. 2002. Indicators for social and economic coping capacity--moving towards a working definition of adaptive capacity. Global Environmental Change 12 (1):25-40.

York, R. 2007. Demographic trends and energy consumption in European Union Nations, 1960-2025. Social Science Research 36 (3):855-872.

Zalasiewicz, J., M. Williams, W. Steffen, and P. Crutzen. 2010. The New World of the Anthropocene. Environmental Science & Technology 44 (7):2228-2231.

Zandbergen, P. A. 2009. Exposure of US counties to Atlantic tropical storms and hurricanes, 1851–2003. Natural Hazards 48 (1):83-99.

Zautra, A., J. Hall, and K. Murray. 2008. Community development and community resilience: an integrative approach. Community Development 39 (3):130-147.

Zellner, M. L., T. L. Theis, A. T. Karunanithi, A. S. Garmestani, and H. Cabezas. 2008. A new framework for urban sustainability assessments: linking complexity, information and policy. Computers, Environment and Urban Systems 32 (6):474-488.

Zhou, L., B. Biswas, T. Bowles, and P. J. Saunders. 2011. Impact of Globalization on Income Distribution Inequality in 60 Countries. Global Economy Journal 11 (1):Article 1.

Zhou, P., B. Ang, and D. Zhou. 2010. Weighting and aggregation in composite indicator construction: a multiplicative optimization approach. Social Indicators Research 96 (1):169-181.

Zhou, P., L.-W. Fan, and D.-Q. Zhou. 2010. Data aggregation in constructing composite indicators: A perspective of information loss. Expert Systems with Applications 37 (1):360-365.

Part 4: Societal Indicator Inventory

Prepared by: Sandra R. Baptista
With input from: Robert S. Chen, David Hastings, Melissa A. Kenney,
Julie Maldonado, and Dale Quattrochi

This section includes a select number of societal indicators that are summarized in 2-3 pages to discuss the specific indicator's uses, data, benefits, and drawbacks. The indicators chosen are those that seemed most relevant to the NCA because of the process used to develop the indicator, the inclusion of climate, the uses of the indicator, or the broad topic. It is not intended to be comprehensive. Part 5 provides a table that includes a more inclusive list of societal indicators and Part 3 includes the list of references inventoried at the time of publication.

Consumer Price Index (CPI)

Approach: Composite index.

Geographic Scope and Scale of Analysis: The Consumer Price Index (CPI) is a fixed-weight index calculated at national, regional, state, and metropolitan area levels by many national statistical agencies. Sub-indices are constructed for categories of goods and services as well as for demographic subgroups, such as the elderly, the poor, and population-size classes.

Users: The CPI is used in several ways:
- as an economic indicator used by business leaders to inform economic decisions and by government officials to inform fiscal and monetary policies and budgetary decisions;
- as a means of adjusting payments to inflation in the public and private sectors (e.g., Social Security benefits, military and Federal Civil Service pension payments, the Food Stamp program, wage increases in collective bargaining agreements, rents, royalties, child support payments, and alimony);
- as a means of adjusting the federal income tax structure (e.g., tax brackets and the standard deduction) to prevent inflation-induced tax changes;
- as a deflator of other economic series, such as the gross domestic product (GDP); and
- by advocacy organizations interested in monitoring changes and trends in compensation inequality, the standard of living, the cost of living, and well-being.

Data Availability: The U.S. Department of Labor's Bureau of Labor Statistics (BLS) publishes the CPI on a monthly basis. Data are available from the BLS Web site: http://www.bls.gov/cpi/. The BLS reports percent change from the previous month, quarter, and year as well as the average change in the prices of consumer goods and services since a base period.

Purposes and Conceptual Framework: The CPI is a measure of the average change over time in the prices of a fixed basket of consumer goods and services that households purchase for day-to-day living each month. The annual or monthly percent change in the CPI provides an estimate of the inflation rate for consumers. The CPI basket of goods and services includes categories such as food and beverages, housing, household furnishings and

operations, apparel, transportation, education and communication, medical care, personal care, and recreation.

Index Composition: Teams of economists, statisticians, computer scientists, and data collectors produce the CPI. In the U.S., the sampling structure used to collect survey data defines: (1) areas or *primary sampling units* (PSUs), (2) a sample of consumer expenditure items, (3) sales outlets and service establishments to be surveyed, and (4) housing units to be surveyed. The expenditure items are classified into categories called *item strata* (for list of categories and items see BLS, 2007). Prices are collected each month in U.S. urban areas, and local-level data are combined to obtain a U.S. city average index.

Origins, Trajectories and Offshoots: The CPI was created in 1919 during World War I to calculate cost-of-living adjustments in wages. It began with studies of family expenditures in 92 industrial centers from 1917 to 1919 and publication of separate indices for 32 cities in 1919. Regular publication of a national index and the U.S. city average index began in 1921 with estimates dating back to 1913.

There are currently three main CPI series:
- CPI for All Urban Consumers (CPI-U);
- Chained CPI for All Urban Consumers (C-CPI-U); and
- CPI for Urban Wage Earners and Clerical Workers (CPI-W).

Since 1978, the BLS has published the CPI-U, which measures the price-change experience of urban consumer units in U.S. Metropolitan Statistical Areas and in urban areas of 2,500 inhabitants or more. It does not cover residents of rural non-metropolitan areas, farm households, military installations, religious communities, or institutions such as prisons and mental hospitals.

The C-CPI-U was introduced in 2002 using data beginning in 2000. It is a chain index, which means that the value of any given period is related to the value of its immediately preceding period, as opposed to a fixed-base index where the value of every period in a time series is directly related to the same value of one fixed-base period. The C-CPI-U is issued at the national level only, first in preliminary form and then subject to two annual revisions. Like the CPI-U, it targets the urban and metropolitan population (roughly 87 percent of the total U.S. population in 1990). Although the C-CPI-U is based on the same prices used to produce the CPI-U and the CPI-W, a different formula and different weights are used to combine basic indices. The C-CPI-U methodology adjusts for consumers' substitutions among expenditure items in reaction to relative price changes thereby accounting for consumers' ability to achieve the same standard of living from alternative sets of goods and services.

The target population of the CPI-W is a subset of the urban population covered by CPI-U and C-CPI-U. The CPI-W is based on the expenditures of urban households for whom 50 percent or more of household income comes from wages earned by hourly wage earning or clerical jobs. To be included in the CPI-W, the household must have at least one earner who has been employed for 37 weeks or more in an eligible occupation during the previous 12 months. In 1990, the Urban Wage Earners and Clerical Workers represented about 32 percent of the total U.S. population.

The CPI methodology has been revised over the years to remove biases that may either overstate or understate the inflation rate. These biases include

- Substitution bias, which means that consumers respond to price changes by shifting their purchases; that is, they tend to substitute lower-priced alternatives for items in the consumer basket that have increased in price;
- Quality bias, which means that over time, technological advances increase the life and usefulness of products, but the CPI does not take these improvements into account;
- New product bias, which means that new products are not introduced into the CPI until they become commonplace, so price decreases associated with the availability of new technologies may not be reflected in the CPI; and
- Outlet bias, which means that the CPI may not adequately reflect consumer shift to new sales outlets such as wholesale clubs and online retailers.

A New CPI for Climate: This index could serve as a model for developing a fixed basket of goods and services organized into major categories containing items for which prices are sensitive to climate-related variables such as temperature (e.g., heating, cooling, and growing degree days), humidity, precipitation, water availability, and frequency and intensity of extreme weather events. This hypothetical Climate CPI and its sub-indices would provide values to inform assessments of climate vulnerability, resilience, and adaptive capacity for U.S. populations in different geographic regions and in different demographic and socioeconomic strata. A Climate CPI might include categories such as food, housing and shelter, health care, transportation access, communication capacity, fresh water supply, and waste water services.

Advantages: Given that the CPI is already widely used and its methodology has been evolving for over 90 years, many key user groups are familiar with the approach, which may facilitate construction, understanding, and adoption of a new Climate CPI.

Drawbacks and Limitations: The CPI reflects the prices of a representative, fixed basket of goods and services purchased by consumers; it does not reflect prices of all goods and services produced and consumed within the country. The CPI is considered a conditional cost-of-living index (COLI) as opposed to an unconditional or complete COLI, which would reflect changes in non-price factors. In other words, unlike a complete COLI, the CPI does not take into account changes in other factors that are known to affect consumer well-being, but are challenging to quantify with a price index, such as security, crime, the value of leisure time, environmental characteristics (e.g., air and water quality), weather conditions, human health, and items provided by governments at no direct cost to consumers. Another challenge for the CPI approach is how to account for changing qualities of commodities and the introduction of new commodities over time.

Sources and further reading:

Bureau of Labor Statistics. Consumer Price Index. Available at http://www.bls.gov/cpi/

Bureau of Labor Statistics. 2007. Chapter 17: The Consumer Price Index. *BLS Handbook of Methods*. Available at http://stats.bls.gov/opub/hom/pdf/homch17.pdf

Greenlees, J. S., & McClelland, R. B. 2008. Addressing misconceptions about the Consumer Price Index. *Monthly Labor Review*, August 2008, 3-19. Available at http://www.bls.gov/opub/mlr/2008/08/art1full.pdf

Ecological Footprint (EF)

Approach: "Systems" or accounting approach.

Geographic Scope and Scale of Analysis: The Ecological Footprint (EF) has been used to assess the sustainability of the global human population, nations, sub-national regions, cities, businesses, organizations, households, individuals, and activities.

Users: Governments, national statistical offices, policy makers, decision makers, businesses, scientists and academics, NGOs, educators and students, and individuals.

Data Availability: Data are available for download in PDF and Excel spreadsheet format at http://www.footprintnetwork.org. The results of the Global Footprint Network's annual National Accounts calculations are presented in the Ecological Footprint Atlas which is available at: http://www.footprintnetwork.org/atlas.

Purposes and Conceptual Framework: The EF is a resource accounting tool for monitoring human demand on the Earth's biosphere and the availability of regenerative and waste absorptive capacity based on prevailing technology and resource management practices. In other words, the EF tool is used to assess whether, and to what extent, ecological assets are being consumed by people either within or beyond the capacity for the regeneration of these assets.

Composition and Methodology: EF analyses calculate how much biologically productive land and water area an activity, individual, organization, business, city, region, or nation demands for resource consumption and waste absorption. This measure is then compared against calculations of biocapacity, i.e., the amount of biologically productive land and water area available to meet these human demands. Analyses include measures of cropland, grazing land, forests, fisheries, the built environment, energy, and biological and non-biological waste. Data are drawn from a variety of sources such as peer-reviewed science journals, thematic collections, the United Nations Statistics Division, the Food and Agriculture Organization of the United Nations, the International Energy Agency, and the Intergovernmental Panel on Climate Change.

The quantity of resources consumed and waste generated in a given year is divided by the yield of the specific land or sea area from which the resources were harvested, or where the waste was absorbed. The result is then converted to global hectares (gha) using yield and equivalence factors. An ecological budget is in balance when the total biocapacity equals the total footprint. If the total footprint exceeds the biocapacity, then the person, population, or activity is operating with an ecological deficit. If the biocapacity exceeds total footprint, then the person, population, or activity has an ecological reserve. The term global ecological overshoot is used to refer to a global ecological deficit. Ecological overshoot leads to the depletion of biological capital (e.g., degraded cropland, diminishing forests, and declining fisheries) and the accumulation of wastes in the biosphere. Overshoot measured on a global scale is used as an indicator of unsustainability. Results at the global scale suggest that humans are consuming resources at a faster rate than the Earth can replenish. Analysts report that, since the mid-1980s, humanity has demanded more regenerative and waste absorptive capacity than the biosphere can supply. In 2007, the estimated world-average EF was 2.7 gha per person (18.0 billion gha total) while the estimated world-average biocapacity was 1.8 gha per person (11.9 billion gha total), suggesting a global ecological deficit of 0.9 gha per person (6.1 billion gha total) (Ewing et al., 2010: 18). This ecological overshoot of 50 percent is also expressed as the need for the equivalent of 1.5 Earths to support worldwide human demand.

Origin, Trajectory, and Offshoots: William Rees and Mathis Wackernagel introduced the EF concept and methodology in the early 1990s and published the book Our Ecological Footprint: Reducing Human Impact on the Earth in 1996. The Global Footprint Network (GFN)—an international nongovernmental organization established in 2003 to advance the science and application of the EF tool—maintains the Ecological Footprint Standards and collaborates with many national governments and international agencies. For instance, in the 2010 Living Planet Report released by the World-Wide Fund for Nature (WWF), the Living Planet Index for monitoring biodiversity is presented along with the EF. In 2004, WWF and BioRegional launched a global initiative to promote the "One Planet Living" framework based on 10 principles of sustainability: (1) zero carbon, (2) zero waste, (3) sustainable transport, (4) sustainable materials, (5) local and sustainable food, (6) sustainable water, (7) land and wildlife, (8) culture and heritage, (9) equity and local economy, and (10) health and happiness.

Advantages: The EF demonstrates a number of desirable qualities:

- the idea of monitoring resource accounts to help balance ecological budgets at different spatial and organizational scales is easily communicated and understandable to policy makers as well as to the general public;
- proponents of the EF method show commitment to an open, transparent, rigorously reviewed scientific process applied in consistent and reproducible ways (e.g., the GFN's development of the Ecological Footprint Standards);
- the use of global hectares (gha) as a common unit of measurement aims to make EF resource accounts results globally comparable and to enable multi-scale and cross-scale analyses of hierarchically nested data;
- calculation of per capita EF allows comparisons of consumption levels and lifestyles; and
- EF resource accounting results are expected to improve as the temporal and spatial resolutions of relevant data sets improve.

Weaknesses and Limitations: The EF method has been criticized for

- inconsistencies in conversions from ha to gha (Wiedmann & Lenzen, 2007);
- working better at international and national levels than at local levels;
- misrepresenting people living in densely populated areas as "parasitic" because they have low levels of biocapacity and must rely on resources imported from other places;
- not taking into account the benefits of trade;
- appearing to reward the replacement of original ecosystems with high-productivity agricultural monocultures by assigning higher biocapacity to such regions;
- not taking into account future technological possibilities and future changes in economic processes;
- inadequately accounting for pollution and toxic control and waste management; and
- not attempting to capture other important aspects of social or economic sustainability such as human health and human well-being.

Sources and further reading:

BioRegional. 2011. One Planet Vision. Available at http://www.oneplanetvision.net

Ewing, B., D. Moore, S. Goldfinger, A. Oursler, A. Reed & M. Wackernagel. 2010. *The Ecological Footprint Atlas 2010*. Oakland, CA: Global Footprint Network.

Global Footprint Network. 2011. Available at http://www.footprintnetwork.org

Kitzes, J. & M. Wackernagel. 2009. "Answers to common questions in Ecological Footprint accounting." *Ecological Indicators*, 9, 812-817.

Wackernagel, M. 2009. "Methodological advancements in footprint analysis." *Ecological Economics*, 68, 1925-1927.

Wackernagel, M. 2011. Global Footprint Network: Our Role in Ending Overshoot. In *Earth Capitalism: Creating a New Civilization through a Responsible Market Economy*, ed. P. U. Petit, 85-90. Piscataway, NJ: Transaction Publishers.

Wackernagel, M. & W. Rees. 1996. *Our Ecological Footprint: Reducing Human Impact on the Earth*. Philadelphia, PA, USA and Gabriol a Island, BC, Canada: New Society Publishers.

Wiedmann, T. & M. Lenzen. 2007. "On the conversion between local and global hectares in Ecological Footprint analysis". *Ecological Economics*, 60, 673-677.

WWF. 2010. Living Planet Report. Available at http://wwf.panda.org/about_our_earth/all_publications/living_planet_report/2010_lpr. WWF & BioRegional. 2011. One Planet Living.
Available at http://www.oneplanetliving.orghttp://www.oneplanetliving.org and http://wwf.panda.org/what_we_do/how_we_work/conservation/one_planet_living.

Environmental Performance Index (EPI)

Approach: Composite index.

Geographic Scope and Scale of Analysis: The Environmental Performance Index (EPI) is designed to inform decision-making and policymaking at global, regional, and national scales. The 2010 EPI ranks 163 countries and groups countries into five regional peer groups: (1) Middle East and North Africa, (2) Eastern Europe and Central Asia, (3) Americas, (4) Europe, and (5) Asia and Pacific. The methodology can also be used to inform environmental protection efforts at state/provincial, local, and corporate scales. Sub-national EPIs have been developed for Abu Dhabi Emirate and China.

Users: Governments, policy makers, decision makers, corporate social responsibility community, environmental scientists, non-governmental organizations (NGOs), advocacy organizations, educators and students, and individuals.

Data Availability: *The 2010 EPI Summary for Policymakers, Main Report, Country Profiles, Indicators Metadata, Sensitivity Analysis,* and data file in Excel spreadsheet format are available for download at http://epi.yale.edu/Fileshttp://epi.yale.edu/Files and http://sedac.ciesin.columbia.edu/es/epi.

Purposes and Conceptual Framework: The EPI is an interdisciplinary information tool designed to facilitate the tracking of environmental performance and to promote decision maker accountability. It is premised on the idea that environmental decision-making and policymaking require robust metrics and can be made more data- and evidence-based. Hence it uses the best available global data sets on environmental performance to measure proximity to established policy targets. Data are drawn primarily from international organizations and research institutions. By providing a baseline for cross-country and cross-sectoral performance comparisons, the EPI helps identify leaders, laggards, and best practices. Comparisons are made issue-by-issue as well as in the aggregate. Using this data-driven approach, the EPI focuses on two overarching policy objectives: (1) reducing environmental stresses on human health and (2) protecting ecosystem vitality.

Methodology: The EPI applies a proximity-to-target methodology to facilitate cross-country comparisons and analysis of how the global community is performing collectively

on each particular policy issue. To construct rankings, raw data are transformed to proximity-to-target scores ranging from the lowest score of zero (worst performance) to the highest score of 100 (at target). The 2010 EPI tracks 25 performance indicators across ten policy categories covering two objectives (Figure 1).

Origin and Trajectory: The EPI originated in 2006, and was based on experience developing the Environmental Sustainability Index (ESI). In part, EPI is a response to the 2000 Millennium Declaration and the Millennium Development Goals. Three versions have been released so far: 2006, 2008, and 2010. The 2012 EPI is currently underway. Over the years, EPI developers at the Yale Center for Environmental Law and Policy and Columbia University's Center for International Earth Science Information Network (CIESIN) have incorporated feedback from many governments and policy makers.

Advantages: EPI efforts have demonstrated the potential for improving environmental performance metrics, refining policy analysis, and understanding the determinants of environmental progress. The EPI helps to identify policy successes, failures, and best practices, and to optimize gains from investments in environmental protection. Among the critical drivers of good environmental results identified are: the level of development, rule of law and good governance, concerted policy effort, and a robust

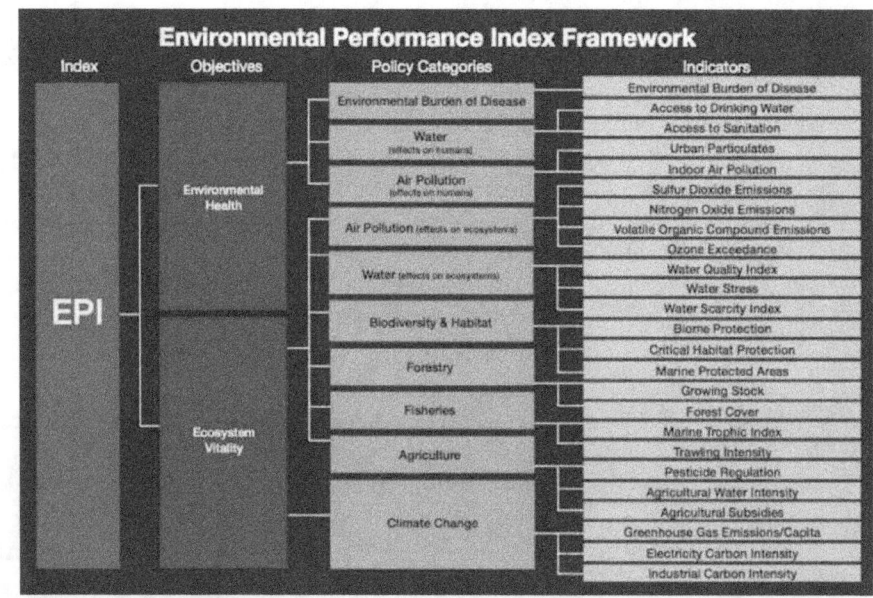

Figure 1. Source: *Emerson, J., D. C. Esty, M.A. Levy, C.H. Kim, V. Mara, A. de Sherbinin, and T. Srebotnjak. 2010. 2010 Environmental Performance Index. New Haven: Yale Center for Environmental Law and Policy.*

regulatory regime. The EPI offers flexibility by using proximity-to-target as the metric and by analyzing performance by specific issue, policy category, peer group, and country. It also provides a model of transparency by providing all of the underlying data online. A subset of the EPI policy issues used for monitoring the trajectory of countries and the global community could be included in a climate change adaptation and mitigation performance measurement system.

Weaknesses and Limitations: EPI authors have been transparent about the limitations of the EPI and its underlying data in order to encourage more rigorous and transparent data collection. They point out that the EPI is hampered by data quality issues for many of the indicators. They call for investments to establish better and broader data collection, methodologically consistent reporting, mechanisms for verification, and a commitment to environmental data transparency. The inability of prior versions to consistently track progress over time has been a major weakness, and in recognition of that fact the 2012 release will focus on a smaller subset of indicators with consistent time series. Simultaneously, with each new release there will be a focused effort to chart a course for improved measurement in one policy category (in 2012 the focus will be on air quality data).

Sources and further reading:

Emerson, J., D. C. Esty, M.A. Levy, C.H. Kim, V. Mara, A. de Sherbinin, and T. Srebotnjak. 2010. *Environmental Performance Index.* New Haven: Yale Center for Environmental Law and Policy.

Environmental Performance Index. 2010. Available at http://epi.yale.edu/ and http://sedac.ciesin.columbia.edu/es/epi

Esty, D. C., Levy, M.A., Kim, C.H., de Sherbinin, A., Srebotnjak, T., & Mara, V. 2008. *Environmental Performance Index.* New Haven: Yale Center for Environmental Law and Policy.

Flood Vulnerability Index (FVI)

Approach: Composite index.

Geographic Scope and Scale of Analysis: The Flood Vulnerability Index (FVI) has been designed for use at three spatial scales: river basin, sub-catchment, and urban area.

Users: Governments, policy makers, decision makers, planners, resource managers, practitioners, NGOs, engineers, researchers, and general public.

Data Availability: The updated methodology and results are disseminated via the FVI Web site at www.unesco-ihe-fvi.org.

Purposes and Conceptual Framework: The FVI is an interdisciplinary tool designed to assess the level of flood vulnerability, at different spatial scales, for three factors: (1) exposure, (2) susceptibility, and (3) resilience. Its purpose is to provide information in support of flood risk management and vulnerability reduction.

Composition and Methodology: Four thematic components are constructed to compute the FVI: (1) social, (2) economic, (3) environmental, and (4) physical (Figure 2). To identify the most significant indicators at each spatial scale, FVI developers have used the derivative method, the correlation method, and a questionnaire to survey expert knowledge. The questionnaire (available at www.unesco-ihe-fvi.org) asks respondents to rate the degree of significance of each indicator, at each spatial scale and for each of the four components, on a scale from 5 (very high influence) to 1 (very low influence). These techniques reduced the number of indicators from an initial set of over 70 candidate indicators to a smaller set of less than 30 indicators. The FVI values range from 0.000 (least vulnerable to flood) to 1.000 (most vulnerable to flood).

Origins and Trajectories: The FVI project Web site, currently hosted by UNESCO-IHE Institute for Water Education, extends previous work by Connor & Hiroki (2005), Balica (2007), Balica et al. (2009), and Balica & Wright (2009, 2010). A Web-based automated tool is being developed to help gather, organize, monitor, process, and compare data for a large number of case studies at various spatial scales. These efforts aim to test the validity of the FVI methodology. The Web interface is being designed to establish a network of knowledge among different institutions and to encourage collaborations. Users are invited to create an account to contribute case study data. In addition to collecting data, the Web-based network of knowledge encourages participants to discuss the concept of vulnerability. In future work, the developers plan to apply the FVI approach to examine coastal cities and the effects of climate change.

Overall Indicators
Relationship between components and factors

Social Component

Exposure	Geographic Scale	Susceptibility	Geographic Scale	Resilience	Geographic Scale
Population density	R,S,U	Past experience	R,S,U	Warning system	R,S,U
Population in flood area	R,S,U	Education (Literacy rate)	R,S,U	Evacuation routes	R,S,U
Closeness to inundation area	R,S,U	Preparedness	R,S,U	Institutional capacity	R,S,U
Population close to coast line	R,S,U	Awareness	R,S,U	Emergency service	R,S,U
Population under poverty	R,S,U	Trust in institutions	R,S,U	Shelters	R,S,U
% of urbanized area	R,S	Communication penetration rate	R,S,U		
Rural population	R,S	Hospitals	R,S,U		
Cadastre survey	S,U	Population with access to sanitation	R,S,U		
Cultural heritage	S,U	Rural population who access to WS	R,S		
% of young & elder	S,U	Quality of Water Supply	S,U		
Slums	U	Quality of Energy Supply	S,U		
		Population growth	S,U		
		Human health	S,U		
		Urban planning	U		

Economic Component

Exposure	Geographic Scale	Susceptibility	Geographic Scale	Resilience	Geographic Scale
Land use	R,S,U	Unemployment	R,S,U	Investment in counter measures	R,S,U
Proximity to river	R,S,U	Income	R,S,U	Infrastructure Management	R,S,U
Closeness to inundation areas	R,S,U	Inequality	R,S,U	Dams & Storage capacity	R,S,U
% of urbanized area	R,S	Quality of infrastructure	R,S,U	Flood insurance	R,S,U
Cadastre survey	S,U	Years of sustaining health life	R,S,U	Recovery Time	R,S,U
		Urban growth	S,U	Past experience	S,U
		Child mortality	S,U	Dikes/ Levees	S,U
		Regional GDP/ capita	S		
		Urban planning	U		

Environmental Component

Exposure	Geographic Scale	Susceptibility	Geographic Scale	Resilience	Geographic Scale
Ground WL	R,S,U	Natural reservations	R,S,U	Recovery time to floods	R,S,U
Land use	R,S,U	Years of sustaining health life	R,S,U	Environmental concern	R,S,U
Over used area	R,S,U	Quality of infrastructure	R,S,U		
Degraded area	R,S,U	Human health	S,U		
Unpopulated land area	R,S	Urban growth	S,U		
Types of vegetation	R,S	Child mortality	S,U		
% of urbanized area	R,S				
Forest change rate	R				

Physical Component

Exposure	Geographic Scale	Susceptibility	Geographic Scale	Resilience	Geographic Scale
Topography (slope)	R,S,U	Buildings Codes	U	Dams & Storage capacity	R,S,U
Geography	R,S,U			Roads	R,S,U
Geology	R,S,U			Dikes / Levees	S,U
Heavy rainfall	R,S,U				
Flood duration	R,S,U				
Return periods	R,S,U				
Proximity to river	R,S,U				
Soil moisture	R,S,U				
Evaporation rate	R,S,U				
Temperature (yearly average)	R,S,U				
River discharge	R,S,U				
Frequency of occurrence	R,S,U				
Flow velocity	S,U				
Storm surge	S,U				
Tidal	S,U				
Flood water depth	S,U				
Sedimentation load	S,U				
Coast line	S,U				
Coastal bathometry	S,U				

Figure 2. Indicators used to compute Flood Vulnerability Indices. Source: UNESCO-IHE (2011)

Advantages: The FVI framework and network of knowledge offer a number of benefits:
- a clear and flexible methodology for stakeholders to evaluate flood vulnerability at multiple spatial scales for different case studies;
- a transparent, collaborative, publicly viewable index development process communicated via a Web site that can continuously identify candidate indicators, test methods for identifying the most significant indicators, collect case study data, and encourage discussions about the concept of vulnerability and how to improve overall index construction;
- the measurement scale from 0.000 (least vulnerable) to 1.000 (most vulnerable) is easily communicated and understood;
- can help raise public awareness of flood vulnerability and climate change risks; and
- can assist governments, decision makers, policy makers, planners, and other stakeholders in setting in priorities, creating coordinated adaptation plans, and promoting resilience.

Weaknesses and Limitations: The following limitations and challenges for FVI development remain:
- the need to strengthen the conceptual framework for FVI construction at multiple spatial scales;
- the need to improve methodologies for identifying the most significant indicators;
- data availability limitations on the occurrence of flooding events and resilience to flood damage;
- insufficiency of the FVI for decision-making, i.e., it should be used in combination with other decision-making tools and approaches such as use of participatory methods, collaboration with multidisciplinary thematic specialists, and consultation with knowledgeable societal representatives; and

- given that the Web site invites users to submit case study data sets, a transparent validation process, that includes users and administrators, is needed in order for each data set entered to be flagged as either validated or not validated.

Sources and further reading:

Balica, S. F. 2007. *Development and Application of Flood Vulnerability Indices for Various Spatial Scales*. MSc, Water Science and Engineering, UNESCO-IHE, Delft.

Balica, S. & N. G. Wright. 2009. "A network of knowledge on applying an indicator-based methodology for minimizing flood vulnerability." *Hydrological Processes*, 23, 2983-2986.

Balica, S. & N. G. Wright. 2010. "Reducing the complexity of the flood vulnerability index." *Environmental Hazards*, 9, 321-339.

Balica, S. F., N. Douben & N. G. Wright. 2009. "Flood vulnerability indices at varying spatial scales." *Water Science & Technology*, 60, 2571-2580.

Connor, R. F. & K. Hiroki. 2005. "Development of a method for assessing flood vulnerability." *Water Science & Technology*, 51, 61-67.

UNESCO-IHE. 2011. Flood Vulnerability Indices (FVI). Available at http://www.unesco-ihe-fvi.org.

Genuine Progress Indicator (GPI) and the Index of Sustainable Economic Welfare (ISEW)

Approach: Accounting approach.

Geographic Scope and Scale of Analysis: The Genuine Progress Indicator (GPI) and its predecessor, the Index of Sustainable Economic Welfare (ISEW), have been designed and applied at national and sub-national levels. In the United States, the GPI has been calculated at the state or sub-state level for California, Maryland, Minnesota, Ohio, Utah, and Vermont. The GPI has been applied in several other countries including Australia, Belgium, Canada, Chile, China, Germany, India, Italy, Japan, Madagascar, Netherlands, New Zealand, Poland, Scotland, Sweden, Thailand, United Kingdom, and Vietnam.

Users: Governments, policy makers, decision makers, NGOs, scientists and academics, and individuals. The GPI has primarily been developed and used by ecological economists.

Data Availability: Data availability differs by country, state, and city. For example, the Utah GPI Calculation Spreadsheet, prepared by the Utah Population and Environment Coalition (UPEC), is available at: http://www.utahpop.org/gpi.html.

Purposes and Conceptual Framework: The GPI and ISEW are approaches for measuring sustainable economic welfare that incorporate social and environmental sustainability components into accounting systems. Created to provide alternatives to Gross Domestic Product (GDP) as a measure of economic performance and social progress, GPI and ISEW account for the costs and benefits of changes in social capital, income distribution, non-market economic activities, environmental conditions, and resource stocks. They can be applied at multiple spatial scales and serve as policy and planning tools for managing progress toward improved societal well-being.

Composition and Methodology: The GPI methodology is based on an accounting framework. Calculation of the GPI begins with the selection of items that account for economic, environmental, and social impacts. The positive and negative values of these components are then summed to obtain a

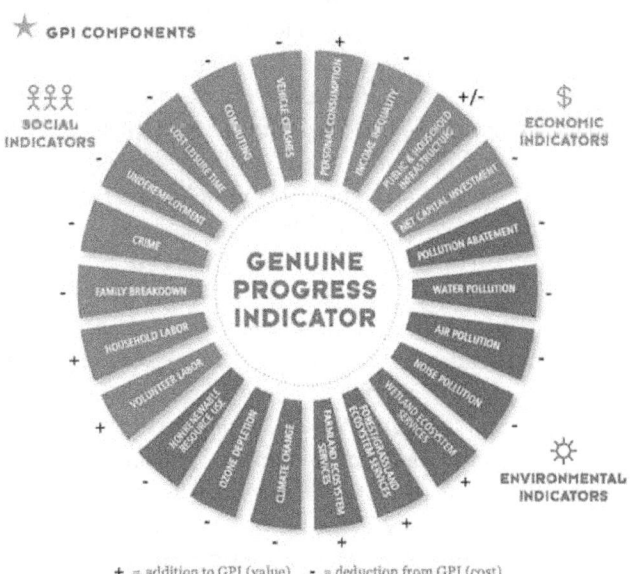

Figure 3. Utah GPI. Source: UPEC (2011), accessed on August 31, 2011.

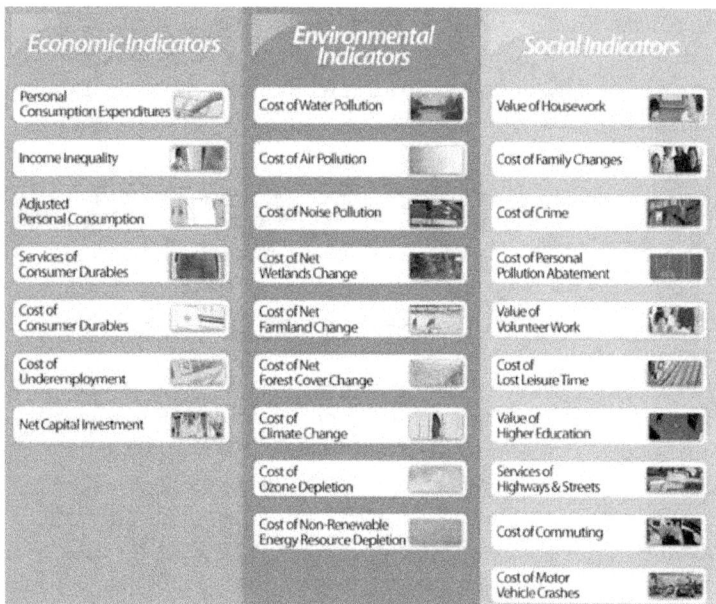

Economic Indicators	Environmental Indicators	Social Indicators
Personal Consumption Expenditures	Cost of Water Pollution	Value of Housework
Income Inequality	Cost of Air Pollution	Cost of Family Changes
Adjusted Personal Consumption	Cost of Noise Pollution	Cost of Crime
Services of Consumer Durables	Cost of Net Wetlands Change	Cost of Personal Pollution Abatement
Cost of Consumer Durables	Cost of Net Farmland Change	Value of Volunteer Work
Cost of Underemployment	Cost of Net Forest Cover Change	Cost of Lost Leisure Time
Net Capital Investment	Cost of Climate Change	Value of Higher Education
	Cost of Ozone Depletion	Services of Highways & Streets
	Cost of Non-Renewable Energy Resource Depletion	Cost of Commuting
		Cost of Motor Vehicle Crashes

Figure 4. Maryland GPI. Source: MD-GPI (2011) http://www.green.maryland.gov/mdgpi/indicators.asp

final index number. The components are expressed in monetary terms, as credits and debits, to facilitate aggregation. Credits typically include the value of personal consumption expenditures, non-market services contributing to welfare (e.g., unpaid household labor, child care, and volunteer work), the services yielded by consumer durables, and road and highway services. Debits include the costs associated with factors such as under-employment, crime and crime prevention, commuting, lost leisure time, divorce, household pollution abatement, motor vehicle accidents, loss of farmland, loss of wetlands, loss of old-growth forests, ozone depletion, resource depletion, air pollution, water pollution, noise pollution, and long-term environmental damage. Some factors, such as net foreign lending/borrowing, may be either positive or negative. Figures 3 and 4 provide examples of GPI composition for the states of Utah and Maryland, respectively.

Origins, Trajectories and Offshoots: The ISEW for the United States was first calculated by economist Herman E. Daly and theologian John B. Cobb, Jr. (Daly & Cobb, 1989). Previous efforts to build a comprehensive indicator of economic welfare include the Measured Economic Welfare (MEW) developed in the early 1970s by Yale University economists William Nordhaus and James Tobin. The MEW was also calculated in monetary terms using personal consumption expenditures as a starting point, then adding or deducting other factors. In 1995, a non-profit research and policy organization, Redefining Progress (based in Oakland, California), began promoting the GPI to replace GDP as a measure of economic and social well-being (Cobb et al., 1995). Since then, many governments, NGOs and others working at country, state, and local levels have developed GPI projects.

Advantages: There is widespread acceptance of the view that the GPI—while still an imperfect measure of progress in need of further refinement—is an improvement over GDP as an indicator of societal well-being. The GPI methodology is highly understandable and can be applied in a consistent and comparable manner across cases and over time allowing for trend analysis at multiple spatial scales. Comparisons of GPI trends can be made across U.S. cities, metropolitan regions, and states. The GPI aggregates positive and negative factors in economic, environmental, and social domains to arrive at a single measure of economic and social well-being. The components are intuitively relevant and the data easily communicated by using monetary values. A version of the GPI can be designed to include the costs and benefits of climate change. Results can serve as warning signs and be used from the national to the local scale to inform debates about development processes as well as to promote accountability of elected officials, policy makers, and decision makers.

Weaknesses and Limitations: The GPI has been criticized for having weak theoretical foundations, inadequately justified key assumptions, danger of indicator bias, and lack of standardization. Some have argued that the GPI is most relevant at the national level of analysis, but is weaker at sub-national or local scales (e.g., Frecker 2005; Clarke & Lawn 2008). Towards achieving international consensus regarding standardization, Kulig & Hoekstra (2010) propose improving the GPI methodology by applying a "hybrid capital approach" in which economic, human, natural, and social capital stocks are distinguished and can be measured in either monetary or non-monetary units.

Sources and further reading:

Bagstad, K. J. & M. Ceroni. 2007. "Opportunities and challenges in applying the Genuine Progress Indicator and Index of Sustainable Economic Welfare at local scales." International Journal of *Environment, Workplace and Employment*, 3, 132-153.

Beça, P., & Santos, R. 2010. "Measuring sustainable welfare: A new approach to the ISEW." *Ecological Economics*, 69(4), 810-819.

Berik, G. N. & E. Gaddis. 2011. The Utah Genuine Progress Indicator (GPI), 1990 to 2007: A Report to the People of Utah. Available at www.utahpop.org/gpi.html.

Bleys, B. 2008. "Proposed changes to the Index of Sustainable Economic Welfare: An application to Belgium." *Ecological Economics*, 64, 741-751.

Castañeda, B. E. 1999. "An index of sustainable economic welfare (ISEW) for Chile." *Ecological Economics*, 28, 231-244.

Clarke, M. & P. Lawn. 2008a. "Is measuring genuine progress at the sub-national level useful?" *Ecological Indicators*, 8, 573-581.

Clarke, M. & P. Lawn. 2008b. A policy analysis of Victoria's Genuine Progress Indictor. *Journal of Socio-Economics*, 37, 864-879.

Cobb, C., T. Halstead & J. Rowe. 1995. If the GDP is up, why is America down? *The Atlantic Monthly*, 276, 59-78.

Costanza, R., J. Erickson, K. Fligger, A. Adams, C. Adams, B. Altschuler, S. Balter, B. Fisher, J. Hike, J. Kelly, T. Kerr, M. McCauley, K. Montone, M. Rauch, K. Schmiedeskamp, D. Saxton, L. Sparacino, W. Tusinski & L. Williams. 2004. Estimates of the Genuine Progress Indicator (GPI) for Vermont, Chittenden County and Burlington, from 1950 to 2000. *Ecological Economics*, 51, 139-155.

Daly, H. E. & J. B. Cobb, Jr. 1989. *For the Common Good: Redirecting the Economy toward Community, the Environment, and a Sustainable Future*. Boston: Beacon Press.

Dietz, S. & E. Neumayer. 2006. Some constructive criticisms of the Index of Sustainable Economic Welfare. In *Sustainable Development Indicators in Ecological Economics*, ed. P. Lawn, 186-206. Cheltenham, UK: Edward Elgar.

England, R. W. 1998. Measurement of social well-being: alternatives to gross domestic product. *Ecological Economics*, 25, 89-103.

Frecker, K. 2005. Beyond GDP: enabling democracy with better measures of social well-being. In *The Kiessling Papers*, ed. J. Willms. Trudeau Centre for Peace and Conflict Studies, University of Toronto.

Gil, S. & J. Sleszynski. 2003. "An index of sustainable economic welfare for Poland." *Sustainable Development*, 11, 47-55.

Hamilton, C. 1999. "The genuine progress indicator methodological developments and results from Australia." *Ecological Economics*, 30, 13-28.

Hanley, N., I. Moffatt, R. Faichney & M. Wilson. 1999. "Measuring sustainability: A time series of alternative indicators for Scotland." *Ecological Economics*, 28, 55-73.

Kulig, A., H. Kolfoort & R. Hoekstra. 2010. "The case for the hybrid capital approach for the measurement of the welfare and sustainability." *Ecological Indicators*, 10, 118-128.

Lawn, P. A. 2003. "A theoretical foundation to support the Index of Sustainable Economic Welfare (ISEW), Genuine Progress Indicator (GPI), and other related indexes." *Ecological Economics*, 44, 105-118.

Lawn, P. & M. Clarke. 2006. Comparing Victoria's Genuine Progress with that of the Rest-of-Australia. *Journal of Economic and Social Policy*, 10, Article 7.

Lawn, P. & M. Clarke. 2010. The end of economic growth? A contracting threshold hypothesis. *Ecological Economics*, 69, 2213-2223.

MD-GPI. 2011. Available at http://www.green.maryland.gov/mdgpi/indicators.asp

Moffatt, I. 1999. "Is Scotland sustainable? A the series of indicators of sustainable development." *International Journal of Sustainable Development and World Ecology*, 6, 242-250.

Moffatt, I. & M. Wilson. 1994. "An Index of Sustainable Economic Welfare for Scotland, 1980-1991." *International Journal of Sustainable Development and World Ecology*, 1, 264-291.

Neumayer, E. 2000. "On the methodology of ISEW, GPI and related measures: some constructive suggestions and some doubt on the 'threshold' hypothesis." Ecological Economics, 34, 347-361.

Nourry, M. 2008. "Measuring sustainable development: Some empirical evidence for France from eight alternative indicators. *Ecological Economics*, 67, 441-456.

Ollivier, T. & P.-N. Giraud. 2010. "The Usefulness of Sustainability Indicators for Policy Making in Developing Countries: The Case of Madagascar." *Journal of Environment & Development*, 19, 399-423.

Talberth, J., C. Cobb & N. Slattery. 2007. The Genuine Progress Indicator 2006: A Tool for Sustainable Development. Oakland, CA: Redefining Progress.

UPEC. 2011. Available at www.utahpop.org/gpi. html. Accessed on August 31, 2011.

Venetoulis, J. & C. Cobb. 2004. The Genuine Progress Indicator 1950-2002 (2004 Update). Oakland, CA: Redefining Progress.

Wen, Z., K. Zhang, B. Du, Y. Li & W. Li. 2007. "Case study on the use of genuine progress indicator to measure urban economic welfare in China." *Ecological Economics*, 63, 463-475.

Human Development Index (HDI)

Approach: Composite index.

Geographic Scope and Scale of Analysis: The Human Development Index (HDI) has been calculated with data collected at the national, state, city, municipal, and village levels.

Users: United Nations Development Programme (UNDP), governments, policy makers, economists, development practitioners, NGOs, businesses, educators and students, individuals.

Purposes and Conceptual Framework: Human development denotes the process of improving human well-being and expanding human freedoms, choices, opportunities and capabilities. The HDI emerged in the 1990s as an alternative to the use of Gross Domestic Product (GDP) alone as a measure for monitoring a country's level of development. The HDI considers three main dimensions of human well-being: health, education, and income. The human development framework has evolved to embrace the themes of environmental sustainability, equity, and empowerment as important aspects of human well-being. Accordingly, in recent years the HDI has been modified and refined by a variety of user communities to analyze human development at sub-national spatial scales and for social groups distinguished by characteristics such as socioeconomic status, gender, age, race and ethnicity.

Index Composition: The HDI is a composite measure of four sub-indicators expressed as values between zero and one. These sub-indicators are: (1) life expectancy at birth, (2) adult literacy, (3) gross school enrollment ratio, and (4) GDP per capita. Based on this composite measure, countries have been ranked and placed into the following categories: Very High Human Development, High Human Development, Medium Human Development, and Low Human Development.

Origins, Trajectories, and Offshoots: The HDI has been reported annually since 1990 in the UNDP Human Development Report (HDR). Pakistani economist Mahbub ul Haq (Feb. 22, 1934 - July 16, 1998) led the team that produced the first HDR and designed the HDI. The Human Poverty Index (HPI) was introduced in the 1997 HDR combining measures of deprivation related to survival, education, and standard of living. The 2010 HDR examined trends and patterns in human well-being since 1970. It found that there are multiple paths to human development. The 2010 HDR introduced a new method of calculating HDI using the Life Expectancy Index (LEI), the Education Index (EI), and the Income Index (II). The 2010 report also presented three new composite indices: Inequality-adjusted HDI (IHDI), a Gender Inequality Index (GII), and a Multidimensional Poverty Index (MPI). Interactive tools at the UNDP Web site http://hdr. undp.org/en/ permit users to build custom indices and explore statistics, graphs, and maps. Equality and sustainability are the central themes selected for the 2011 HDR due for release in November 2011.

The American Human Development Project (AHDP) of the **Social Science Research Council** (SSRC) introduced the American Human Development Index (American HD Index) in a report titled *The Measure of America: American Human Development Report 2008-2009* modeled on the UNDP HDR. The American HD Index uses official government data to create a composite rating of overall well-being based on health, education, and income indicators. An updated report titled *The Measure of America 2010-2011: Mapping Risks and Resilience* was released in November 2010. The American HD Index allows for well-being rankings of the 50 states, 435 congressional districts, county groups within states, major metropolitan areas, women and men, and racial and ethnic groups. In 2009 the AHDP released reports focused on the states of Louisiana and Mississippi. In 2010 it released *A Century Apart: New Measures of Well-being for U.S. Racial and Ethnic Groups*.

Linkages of HDI to Climate Change: Inaction in the face of climate change could derail decades of progress in human development. The adverse impacts of climate change could, for example, contribute to water scarcity, food insecurity, disaster risk, and migration. Such impacts could have negative effects on health, education, income, and other indicators of human well-being. Sustainable human development requires improved understanding of the linkages between economic activities, greenhouse gas emissions, and the social-ecological systems supplying energy, raw materials, infrastructure, shelter, water, and food.

Pros and Cons of HDI: Introduction of the HDI about two decades ago was successful in encouraging analysts to measure human development and well-being in new ways by combining health and education metrics with economic parameters such as GDP. The HDI served as a straightforward and manageable index and was widely accepted by a variety of users. Earlier efforts to calculate the HDI failed to adequately consider equity, sustainability, or ecological dimensions. Initially, analyses were limited to the country-level and reported values for heterogeneous populations using averages. Increasingly, however, analysts have been modifying the original HDI approach to consider additional components of human well-being related to inequality, poverty, gender, sustainability, human security, empowerment, and governance. Advances in data availability and geospatial technologies enable analysts to examine inequalities of human development at finer spatial scales (e.g., within countries, states, or counties) and over longer temporal scales to help reveal how changes along different dimensions of well-being are socially and spatially distributed. These methodological advances should improve the generation of knowledge needed to inform policy.

Sources and further reading:

Moran, D. D., Wackernagel, M., Kitzes, J. A., Goldfinger, S. H., & Boutaud, A. 2008. Measuring sustainable development — Nation by nation. *Ecological Economics*, 64(3), 470-474.

SSRC. 2011. American Human Development Project._http://measureofamerica.org/

UNDP. 1990-2011. Global Human Development Reports._http://hdr.undp.org/en/reports/

UNDP. 2009. *Linking climate change policies to human development analysis and advocacy: a guidance note for Human Development Report teams.* United Nations Development Programme.

UNDP. 2010. Human Development Report 2010. *The Real Wealth of Nations: Pathways to Human Development.* Available at http://hdr.undp.org/en/reports/global/hdr2010/

Human Security Index (HSI)

Approach: Composite index.

Geographic Scope and Scale of Analysis: The Human Security Index (HSI) is being developed for use at national and sub-national levels. The current global version (HSIv2) covers 232 countries. Development of a county-level prototype HSI for the United States (HSI USA) is underway.

Users: Governments, policy makers, decision makers, planners and managers of public and private services, NGOs, data and indicator developers, remote sensing and GIS specialists, scientists, researchers, academics, educators, students, communities, and individuals.

Data Availability: Data in Excel and ISO Open Document formats, sample maps, and documentation of indicator data sources and references are available online at www.humansecurityindex.org, which is hosted by researchers at Osaka City University.

Purposes and Conceptual Framework: The HSI is a tool designed to characterize economic, environmental, and social security at spatial scales ranging from national to local for the purpose of guiding strategies to improve place-based conditions related to these three themes. It has been conceptualized as a more thematically comprehensive and geographically extensive measure than the Human Development Index (HDI).

Composition and Methodology: Three sub-indices are constructed to compute the HSI: (1) the Economic Fabric Index, (2) the Environmental Fabric Index, and (3) the Social Fabric Index. Together these three thematic components incorporate over 30 "leading" indicators, including some composite indicators. Thus, overall, the global HSIv2 incorporates over 150 input

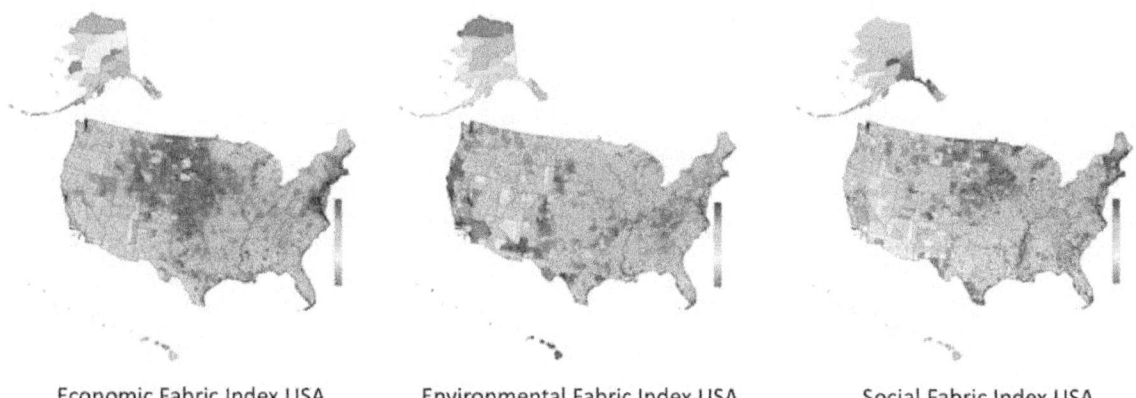

Economic Fabric Index USA   Environmental Fabric Index USA   Social Fabric Index USA

Figure 5. Maps displaying the Economic, Environmental, and Social Fabric Indices for the United States (Prototypes, Version 0.1). Blue indicates relatively better situations. Source: Hastings (2011a).

data sets. The Economic Fabric Index aims to characterize financial resources, economic (in) equality, and financial-economic governance. The Environmental Fabric Index integrates data related to environmental living conditions, environmental protection and governance, disaster risk and vulnerability, environmental sustainability, and population change. The global Social Fabric Index consists of six subcomponents: (1) education and information empowerment, (2) diversity, (3) peacefulness, (4) food security, (5) health, and (6) governance. The prototype Social Fabric Index for the United States currently has four subcomponents:

(1) education, (2) health, (3) crime and punishment, and (4) social stress. As in the HDI, values are scaled from 0.000 (low human security) to 1.000 (high human security). Both the global HSIv2 and HSI USA have been computed by averaging scaled input data with equal weights into subcomponents, averaging the subcomponents into their respective Fabric Indices, and then averaging the three Fabric Indices with equal weighting into the HSI (Hastings, 2011a).

Origins and Trajectories: David A. Hastings introduced a prototype of the global HSI in December 2008 encompassing 200 nation-societies (Hastings. 2008). The improved global HSIv2, covering 232 national-level societies, was released in December 2010 (Hastings, 2010b). Formulation of the HSI USA was initiated in 2009, resulting in county-level prototypes since July 2010 (Figures 5 and 6). The HSI USA currently incorporates about 35 indicators, including composite indicators such as a Natural Amenity Index (modified from the U.S. Department of Agriculture's Economic Research Service) and the Healthy Food Access Index (U.S. Department of Health and Human Service). Work has begun on applying

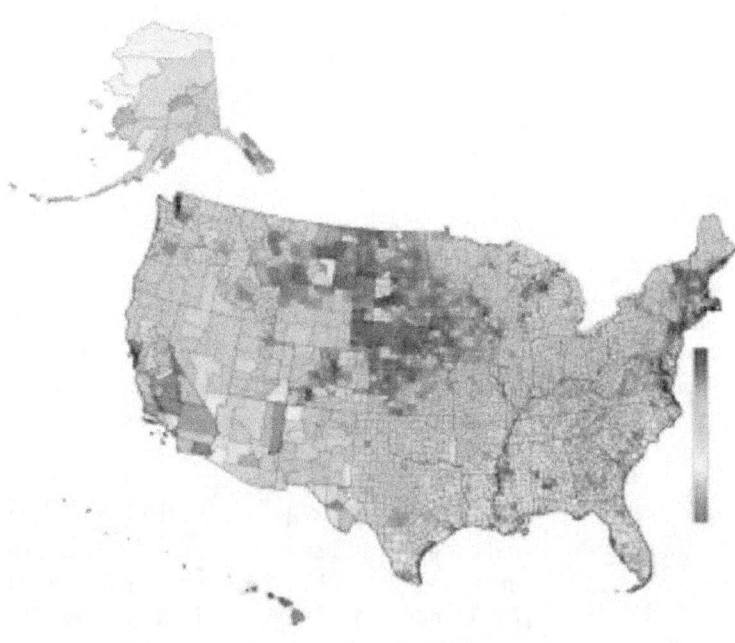

Figure 6. Map displaying the Human Security Index for the United States (HSI USA Prototype, Version 0.1). Blue indicates relatively better situations. Source: Hastings (2011a).

the HSI approach to developing countries (e.g., Thailand) at sub-national levels.

Advantages: The HSI framework offers several benefits:

- the flexibility to explore the integration of candidate indicators/indices on a diversity of topics related to overall human security, including the human security dimensions of climate change and variability (e.g., climate-related public health issues and the vulnerability of coastal communities);
- thematic indicators/indices can be combined at multiple spatial scales using a nested approach;
- a working platform for individuals and teams to identify existing indicators/indices, to strengthen thematic sub-indices as new data sets or indicators emerge, and to contribute to the improvement of data collection systems;
- a platform for working towards a globally harmonized conceptualization and methodology to measure human (in)security; and
- a transparent, publicly viewable index development process communicated via a Web site that, in addition to presenting data and maps, aims to foster discussions about human security themes, indicators, sub-indices, and how to improve overall index construction.

Weaknesses and Limitations: The following challenges and needs for HSI development remain:

- strengthen the conceptual framework for HSI construction at multiple spatial scales;
- continue to address data availability limitations (it is an ongoing challenge to find, improve, and create appropriate source data sets as well as documentation);
- improve methodologies to assess candidate input indicators/indices;
- improve use at higher resolutions (e.g., at the community level);
- better engagement with data/indicator developers and users for mutual benefit; and
- encourage greater input from the Earth observation and other professional communities.

Sources and further reading:

Hastings, D. A. 2008. Describing the human condition – from human development to human security: an environmental remote sensing and GIS approach. *GIS-IDEAS 2008 Conference*, Hanoi, Vietnam, December 4-6, 2008. Available at http://wgrass.media.osaka-cu.ac.jp/gisideas08/viewpaper.php?id=299.

Hastings, D. A. 2009. From Human Development to Human Security: A Prototype Human Security Index. UNESCAP Working Paper WP/09/03, Macroeconomic Policy and Development Division. Bangkok, Thailand: United Nations Economic and Social Commission for Asia and the Pacific. Available at http://www.unescap.org/pdd/publications/workingpaper/wp_09_03.pdf.

Hastings, D. A. 2010a. The global human security index: Can disaggregations help us to forge progress? *The Coastal Society's 22nd International Conference, Shifting Shorelines: Adapting to the Future*, Wilmington, N.C., June 13-16, 2010. Available at http://nsgl.gso.uri.edu/coastalsociety/TCS22/papers/Hastings_2_papers.pdf.

Hastings, D. A. 2010b. The Human Security Index: An update and a new release. *GIS-IDEAS 2010 Conference*, Hanoi, Vietnam, December 9-10, 2010. Available at http://wgrass.media.osaka-cu.ac.jp/gisideas10/viewabstract.php?id=381.

Hastings, D. A. 2011a. The Human Security Index: Potential Roles for the Environmental and Earth Observation Communities. *Earthzine*. Available at http://www.earthzine.org/2011/05/04/the-human-security-index-potential-roles-for-the-environmental-and-earth-observation-communities_

Human Security Index. 2011b. Available at http://www.humansecurityindex.org._

Quantifying Vulnerability to Climate Change: Implications for Adaptation Assistance

Approach: Composite index.

Geographic Scope and Scale of Analysis: The current scope is global, covering 233 countries.

Users: Governments, NGOs, policy makers, decision makers, planners, scientists and academics, educators, students, and individuals. A major focus is on economic impacts/costs with respect to possible rectification mechanisms, including aid organizations.

Data Availability: Wheeler (2011b) is a spreadsheet with digested data.

Purposes and Conceptual Framework: A set of risk indicators has been crafted for three problem areas: increasing frequency of weather-related disasters, sea level rise, and loss of agricultural productivity. For each of these arenas, indicators enumerate: (1) climate drivers, (2) climate vulnerability, considering income and regulation, (3) project concerns, considering project cost and probability of success, (4) population, (5) climate drivers indicated aid share, using climate drivers adjusted by population, (6) climate vulnerability indicated aid share, using climate vulnerability adjusted by population, and (7) project concerns indicated aid share, using project concerns adjusted by population.

Methodology and Composition: The paper embeds these indicators in a methodology for cost-effective allocation of adaptation assistance. The methodology can be applied easily and consistently to all 233 countries and all three problems, or to any subset that may be of interest to particular donors. Institutional perspectives and priorities differ; the paper develops resource allocation formulas for three cases: (1) potential climate impacts alone, as measured by the three indicators; (2) case 1 adjusted for differential country vulnerability, which is affected by economic development and governance; and (3) case 2 adjusted for donor concerns related to project economics: inter-country differences in project unit costs and probabilities of project success. The paper is accompanied by an Excel database with complete data for all 233 countries. It provides two illustrative applications of the database and methodology: assistance for adaptation to sea level rise by the 20 island states that are both small and poor and general assistance to all low-income countries for adaptation to extreme weather changes, sea-level rise, and agricultural productivity loss.

Note that the methodology involves an attempt to separate the effects of climate change, income and governance, and estimating the effect of the latter two variables on vulnerability to climate change.

Origins, Trajectories and Offshoots: With respect to sea level rise, the foundation is previous work for a subset of developing countries (Dasgupta et al., 2009a,b). The current effort extends coverage to a full set of coastal and island states. Similarly, the agricultural productivity exercise extends the groundbreaking work of Cline (2007) to the full set of 233 countries.

Strengths and Advantages: The approach is designed to be thematically and geographically scalable. (Thus, though focused on global situations, particularly with respect to potential assistance by aid donors, it should be adaptable to state or county level in the U.S., related to possible economic/fiscal vulnerabilities related to disaggregated aspects of climate change and needs/opportunities for diverse actors to prepare and mitigate such vulnerabilities.)

Weaknesses and Limitations: Currently focused on global situations, adaptation to community levels within the U.S. would take some work – including developing appropriate source data. (The editor of this summary imagines that such adaptation could be done, through (1) a compilation of event-based data, and (2) an adaptation of climate models ported into county-based model summaries.) So this may not be a weakness or limitation in the current work – so much as a challenge in adapting the approach to internal sub-national application.

Sources and further reading:

Cline, William. 2007. Global Warming and Agriculture: Impact Estimates by Country. Washington, DC: Center for Global Development and Peterson Institute for International Economics.

Dasgupta, Susmita, Benoit Laplante, Craig Meisner, David Wheeler and Jianping Yan. 2009a. The Impact of Sea Level Rise on Developing Countries: a Comparative Analysis. Climatic Change, 93:379–388.

Dasgupta , Susmita, Benoit Laplante, Siobhan Murray and David Wheeler. 2009b. Climate Change and the Future Impacts of Storm surge Disasters in Developing Countries. Center for Global Development Working Paper No. 182. http://www.cgdev.org/content/publications/detail/1422836

EM-DAT. 2010. The International Disaster Database. Center for Research on the Epidemiology of Disasters. http://www.emdat.be/

Wheeler, David, 2011a. Quantifying Vulnerability to Climate Change: Implications for Adaptation Assistance. Center for Global Development Working Paper 240. 53 pp. http://www.cgdev.org/files/1424759_file_

Wheeler_Quantifying_Vulnerability_
FINAL.pdf

Wheeler, David, 2011b. Vulnerability to Climate Change. (Data base in spreadsheet form covering 233 countries associated with Wheeler, 2011A) Center for Global Development. http://www.cgdev.org/files/1424986_file_Quantifying_Vulnerability_DB.xls

Two Web links which may serve as jumping-off points are:
http://www.cgdev.org/content/publications/detail/1424759/?utm_&&&
http://www.cgdev.org/content/publications/detail/1424986

Social Vulnerability Index (SoVI)

Approach: Composite index.

Geographic Scope and Scale of Analysis: The Social Vulnerability Index (SoVI) for the United States has been calculated at the county, city, census tract, and census block group levels (Cutter et al., 2003; Cutter & Finch, 2008; Schmidtlein et al., 2008). Adaptations of the SoVI approach have been applied at the municipal level in Portugal (Mendes, 2009) and Norway (Holand et al., 2011).

Users: Governments, NGOs, policy makers, decision makers, planners, scientists and academics, educators, students, and individuals.

Data Availability: The data used to construct the SoVI for the U.S. are drawn from the U.S. Census Bureau and other national data sources. Maps and data are available at the University of South Carolina's Hazards and Vulnerability Research Institute (HVRI) Web site: http://webra.cas.sc.edu/hvri/products/sovi_data.aspx

Purposes and Conceptual Framework: The SoVI provides a comparative metric of social vulnerability to environmental hazards. It supports an integrative vulnerability science approach to hazards research. Extending Cutter's (1996) hazards-of-place model of vulnerability, which integrates physical and social factors, the conceptual basis of the SoVI is that social vulnerability is multidimensional and dynamic. The approach recognizes that the ability of communities and individuals to respond to, cope with, recover from, and adapt to environmental hazards is influenced by social, economic, demographic, built environment, and housing

characteristics. Time-series maps of SoVI results help reveal patterns of geographic variation in social vulnerability to environmental hazards and disaster recovery.

Methodology and Composition: The current version of the SoVI for U.S. counties synthesizes 32 variables derived from the research literature on hazard impacts and disaster preparedness, response, and recovery. The data are standardized and, using a principal component analysis (PCA), reduced into a smaller set of key factors of vulnerability. These key components are summed to arrive at a single numerical value that represents the social vulnerability for each county, and these composite scores are displayed in relation to each other.

Origins, Trajectories and Offshoots: Originally developed by Cutter et al. (2003), the first version of the SoVI employed 42 socioeconomic, demographic, and built environment variables to examine social vulnerability for all 3,141 U.S. counties in 1990. Initially, 250 variables were selected for consideration because they corresponded to the social characteristics identified in the hazards research literature as contributing to vulnerability. This pool was then reduced to a smaller set of 42 independent variables, which were normalized to a fixed scale (percentages, per capita, or per square mile). Eleven components were selected by performing a PCA. All factors were given equal importance. These 11 key factors explained about 76% of the total variation among U.S. counties, broken down as follows: personal wealth (12.4% of the variation), age (11.9%), density of the built environment (11.2%), single-sector economic dependence (8.6%), housing stock and tenancy (7.0%), race—African American (6.9%), ethnicity—Hispanic (4.2%), ethnicity—Native American (4.1%), race—Asian (3.9%), occupation (3.2%), and infrastructure dependence (2.9%). Subsequent calculations of the SoVI for the U.S. resulted in a total of 11 to 12 components explaining 73% to 78% of the overall variation among U.S. counties in 1960, 1970, 1980, 1990, and 2000 (Cutter & Finch, 2008).

Schmidtlein et al. (2008) assessed the sensitivity of the SoVI by studying the impacts of changes in scale, changes in variable selection, and differences in geographic context. They applied SoVI to three study areas at the census tract level: Charleston, SC; Los Angeles, CA; and New Orleans, LA. They found the SoVI algorithm to be fairly robust to minor changes in variable selection and to downscaling from the county to census tract level. However, the algorithm's sensitivity to changes in index construction varied across study areas.

Oxfam America commissioned Susan Cutter and Christopher Emrich, at the University of South Carolina's HVRI, to apply SoVI to climate change-related hazards (Oxfam America, 2009). This commissioned study focused on the 13-state region of the U.S. Southeast (Alabama, Arkansas, Florida, Georgia, Kentucky, Louisiana, Maryland, Mississippi, North Carolina, South Carolina, Tennessee, Texas, and Virginia), which encompasses roughly 80% of all U.S. counties characterized by persistent poverty. This project (SoVI-SE) used 32 variables to define the multiple dimensions of vulnerability. The following eight components accounted for most of the variation in social vulnerability: wealth, age, race, gender, ethnicity, rural farm population, special needs populations, and employment status. The report, titled "Exposed: social vulnerability and climate change in the US Southeast," is available at http://adapt.oxfamamerica.org/.

Strengths and Advantages: The SoVI approach offers a useful methodology for quantifying spatial and temporal variations in the relative levels of social vulnerability to environmental hazards as well as a tool for modeling scenarios of potential future vulnerabilities. It can be applied to specific areas of interest that are expected to be most impacted by climate change, such as coastal, riverine, or dryland counties, cities, census tracts, or census block groups. SoVI calculations can be analyzed with hazard event frequency and economic loss data for specific hazard types or by specific time periods for multiple hazards.

Weaknesses and Limitations: As with many other indices, assessment of the conceptual, theoretical, and methodological validity of the SoVI remains a challenge. Data availability is another important constraint. Future work should continue to address the various subjective decisions made in the index construction process and to explore methodologies for determining relative weights. Past efforts have lacked a sufficient theoretical basis for making reliable judgments about the relative importance of index components, and have therefore weighted factors equally to arrive at composite SoVI scores.

Sources and further reading:

Cutter, S. L., & Finch, C. 2008. "Temporal and spatial changes in social vulnerability to natural hazards." *Proceedings of the National Academy of Sciences*, 105(7), 2301-2306.

Cutter, S. L., B. J. Boruff & W. L. Shirley. 2003. "Social vulnerability to environmental hazards." *Social Science Quarterly*, 84(2), 242-261.

Hazards and Vulnerability Research Institute (HVRI). Social Vulnerability Index for the United States. University of South Carolina. Available at http://www.sovius.org or http://webra.cas.sc.edu/hvri/products/sovi.aspx

Holand, I. S., P. Lujala & J. K. Rød. 2011. "Social vulnerability assessment for Norway: A quantitative approach." *Norsk Geografisk Tidsskrift* - Norwegian Journal of Geography, 65, 1-17.

Mendes, J. M. d. O. 2009. "Social vulnerability indexes as planning tools: beyond the preparedness paradigm." *Journal of Risk Research*, 12, 43-58.

Oxfam America. 2009. Exposed: social vulnerability and climate change in the US Southeast. Boston, MA: Oxfam America Inc. Available at http://www.oxfamamerica.org/files/Exposed-Social-Vulnerability-and-Climate-Change-in-the-US-Southeast.pdf

Schmidtlein, M. C., R. C. Deutsch, W. W. Piegorsch, S. L. Cutter. 2008. "A sensitivity analysis of the Social Vulnerability Index." *Risk Analysis* 28(4), 1099-1114.

Sustainable Society Index (SSI)

Approach: Composite index.

Geographic Scope and Scale of Analysis: The Sustainable Society Index (SSI) has been applied at global, regional, national, and sub-national levels. Its most recent edition assesses the level of sustainability of 151 countries.

Users: Governments, policy makers, decision makers, NGOs, businesses, scientists and academics, educators and students, and individuals.

Data Availability: The input data and the scores for the indicators, categories, well-being dimensions, and the overall index of the three editions of the

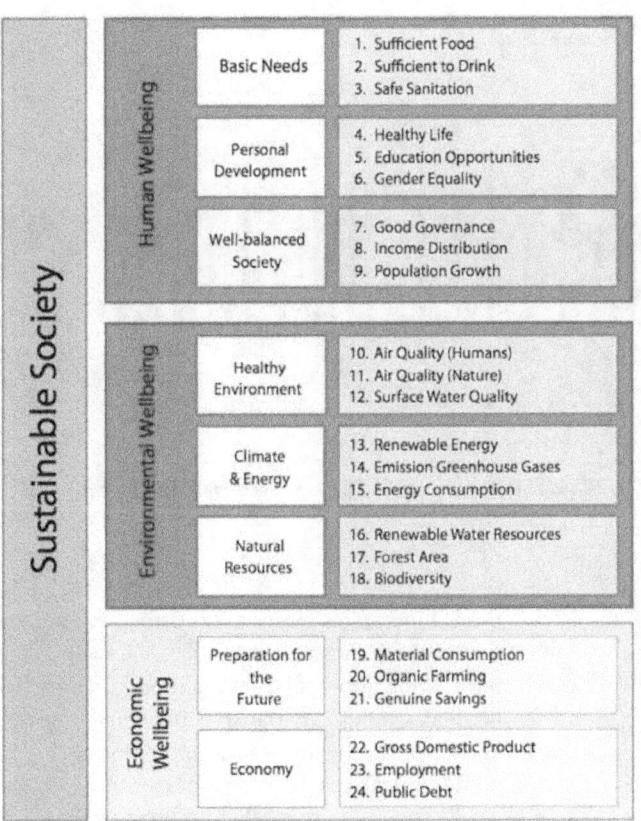

Figure 7. The SSI-2010 framework. Source: http://www.ssfindex.com/ssi/framework

SSI (2006, 2008, and 2010) can be downloaded in Excel format from the Sustainable Society Foundation (SSF) Web site: http://www.ssfindex.com/data. The site also provides interactive maps to visualize country-level scores for all three editions: http://www.ssfindex.com/maps.

Purposes and Conceptual Framework: The SSI is a tool designed to measure and monitor levels of sustainability. Its goal is to provide a simple, transparent, and easily understandable integrated set of sustainability and quality of life indicators.

http://www.ssfindex.com/ssi/framework

Composition and Methodology: The SSI-2010 framework includes 24 indicators organized into eight categories covering three dimensions of well-being (Figure

7). All indicators, categories, well-being dimensions, and the overall index are scored on a scale from 0 to 10, where the target sustainability value is 10. Full sustainability is achieved when the sustainability value for all 24 indicators is 10. The overall SSI country-level scores are calculated as the unweighted average of the 24 indicators, and the overall global scores are calculated as the unweighted average of the 151 countries. According to the SSI-2010, the United States ranks 50th with a score of 6.21, while Switzerland ranks 1st with a score of 7.55 and Sudan ranks 151st with a score of 4.54. The overall global score of the SSI increased slightly from 5.76 in 2006 to 5.92 in 2008 to 5.94 in 2010 (see Figures 8 and 10). The global score of the Climate and Energy category decreased over the period 2006 to 2010 (Figures 9 and 10).

Origins and Trajectories: The SSF was established in 2006 by Geurt van de Kerk and Arthur Manuel as a private initiative to develop the SSI and to publish and disseminate results every two years. The first two editions of the SSI were published in December 2006 (150 countries) and December 2008 (151 countries). These editions were based on a framework of 22 indicators organized into five categories

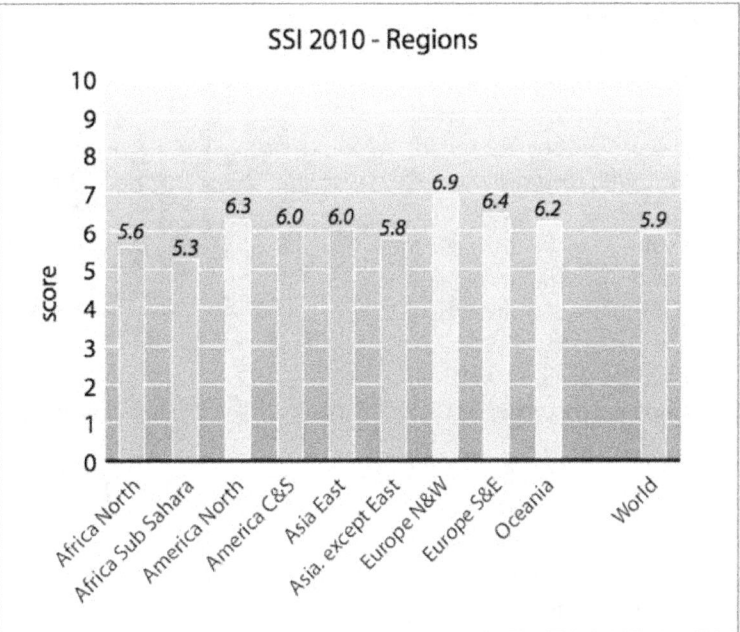

Figure 8. The SSI-2010 regional and global scores. Source: van de Kerk & Manuel (2010:15)

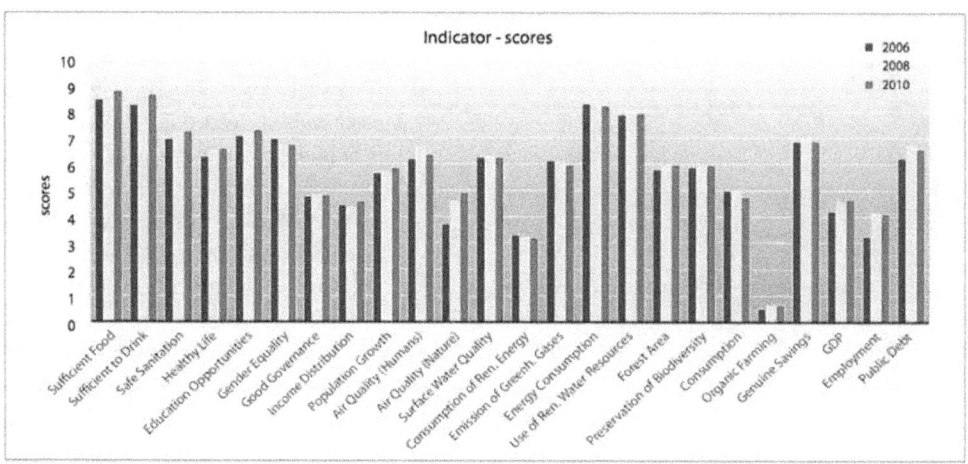

Figure 9. Indicator scores for the global SSI-2006, SSI-2008, and SSI-2010. Source: van de Kerk & Manuel (2010:17)

(personal development, healthy environment, well-balanced society, sustainable use of resources, and sustainable world). The categories were derived from a comprehensive definition of a sustainable society that van de Kerk and Manuel refer to as "the Brundtland+ definition," which is the definition of the 1987 report of the World Commission on Environment and Development (Brundtland Commission) plus explicit inclusion of the social aspects of human life: "A sustainable society is a society that meets the needs of the present generation, that does not compromise the ability of future generations to meet their own needs, in which each human being has the opportunity to develop itself in freedom, within a well-balanced society and in harmony with its surroundings" (van de Kerk & Manuel 2010:12). In an effort to make the indicator framework more balanced and transparent, the SSI was evaluated and redesigned for the 2010 update (third edition), which was published in December 2010.

Advantages: The SSI framework offers the following advantages:
• input data are collected from public sources such as scientific institutes and international organizations;

• the SSI integrates existing indicators, including composite indicators;
• the scoring approach is straightforward and facilitates quick comparisons between regions and countries, using graphs and maps to communicate results at a glance;
• regular updates are made to monitor trends (results are published and disseminated every two years);

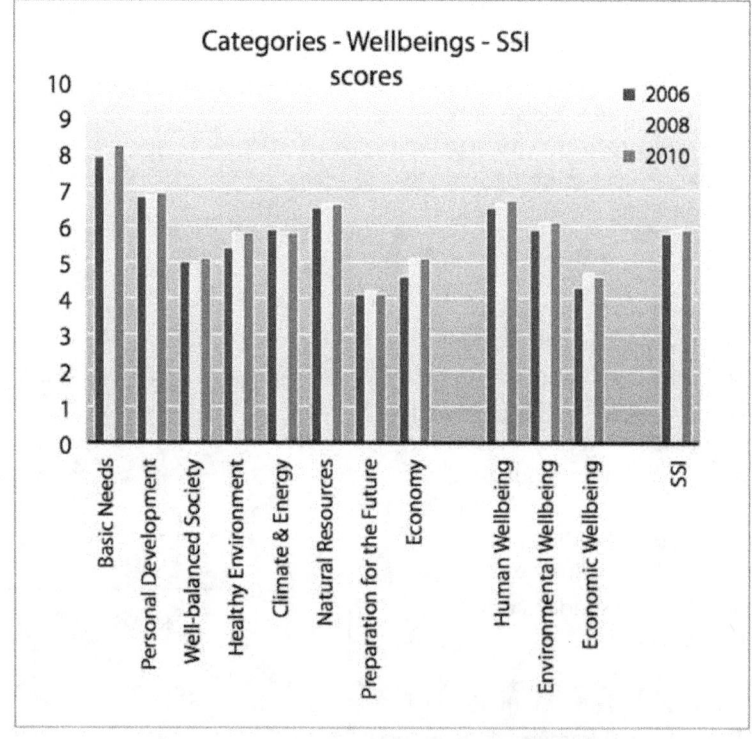

Figure 10. Category, well-being dimension, and overall scores for the global SSI-2006, SSI-2008, and SSI-2010. Source: van de Kerk & Manuel (2010:18)

- the effort draws on the expertise of a worldwide network of contributors; and
- the SSF Web site offers an easy-to-use interactive map to explore the data.

Weaknesses and Limitations: The following needs for further SSI development remain:
- address data availability limitations (e.g., 43 countries were left out of the analysis due to lack of data);
- address concerns regarding the reliability of the available input data;
- strengthen methodologies used to assess candidate input indicators/indices; and
- improve use at higher resolutions (e.g., at sub-national levels).

Sources and further reading:

van de Kerk, G. & A. R. Manuel. 2008. "A comprehensive index for a sustainable society: the SSI — the Sustainable Society Index." *Ecological Economics*, 66, 228-242.

van de Kerk, G. & A. R. Manuel. 2009. Sustainable Society Index. In *Encyclopedia of Earth*. Eds. Cutler J. Cleveland (Washington, D.C.: Environmental Information Coalition, National Council for Science and the Environment). Available at http://www. eoearth.org/article/Sustainable_Society_ Index.

van de Kerk, G. & A. Manuel. 2010. *Sustainable Society Index* 2010. Sustainable Society Foundation. Available at http://www. ssfindex.com/information/publications.

Sustainable Society Foundation. 2011. Available at http://www.ssfindex.com.

Part 5: Societal Indicator Inventory Table

Prepared by: Sandra R. Baptista
With input from: Robert S. Chen, David Hastings, Melissa A. Kenney,
Julie Maldonado, and Dale Quattrochi

#	Name of Indicator (alphabetical order)	Indicator Approach	Organizations/ Authors/Contacts	Relevant Web Sites	Year Initiated
1.	Agenda 21 Indicators (UNCSD Indicators of Sustainable Development)	basket of 134 indicators	United Nations, Commission on Sustainable Development (UNCSD)	http://www.un.org/esa/dsd/agenda21/res_agenda21_40.shtml http://www.un.org/esa/dsd/dsd_aofw_ind/ind_index.shtml http://www.un.org/esa/dsd/dsd_aofw_ind/ind_csdindi.shtml http://www.un.org/esa/sustdev/natlinfo/indicators/isdms2001/table_4.htm	1992
2.	Albuquerque Progress Report (APR)	basket of 62 indicators (8 City Goals); Desired Community Conditions & Community Report Cards	Albuquerque Indicators Progress Commission (AIPC), City of Albuquerque, NM	http://www.cabq.gov/progress	1998
3.	Arctic Water Resources Vulnerability Index (AWRI)	composite water index (27 indicators)	Resilience and Adaptive Management Group, University of Alaska Anchorage; Lilian Alessa and Andrew Kliskey	http://ram.uaa.alaska.edu/AWRVI.htm	2007
4.	Arizona Indicators Project	basket of indicators	Managed by Morrison Institute for Public Policy, Arizona State University; Pat Gober	http://arizonaindicators.org	2007
5.	Asset Index (U.S. states)	basket of 39 indicators; comparative ranking of U.S. states from 1 to 50 (best to worst)	Asset Development Institute, Center on Hunger and Poverty, Heller Graduate School for Social Policy and Management, Brandeis University	http://iasp.brandeis.edu/pdfs/assetindex.pdf	2002
6.	Baseline Resilience Index for Communities (BRIC)	composite index (disaster resilience)	Community and Regional Resilience Institute (CARRI); Susan Cutter, C. Burton and C. Emrich, University of South Carolina	http://www.resilientus.org	2010
7.	Basic Capabilities Index (BCI)	composite index	Social Watch; previously "Quality of Life Index"	http://www.socialwatch.org/node/9240 http://www.socialwatch.org/node/11389	2005
8.	Beach Tourism Vulnerability Index (BTVI)	composite index (vulnerability to climate change)	Sabine Perch-Nielsen	http://dx.doi.org/10.1007/s10584-009-9692-1 http://dx.doi.org/10.1007/s10584-009-9692-1	2010
9.	Boston Indicators Project	basket of indicators	The Boston Foundation in partnership with the City of Boston and the Metropolitan Area Planning Council	http://www.bostonindicators.org	2000
10.	Built Environment Vulnerability Index (BEVI)	composite index (vulnerability to natural hazards)	Kevin Borden (University of South Carolina) et al.	http://www.bepress.com/jhsem/vol4/iss2/5	2007
11.	Canadian Water Sustainability Index (CWSI)	composite index (fresh water and community level wellbeing)	Policy Research Initiative	http://www.policyresearch.gc.ca/doclib/PR_SD_CWSI_200702_e.pdf	2005
12.	Central Texas Sustainability Indicators Project (CTSIP)	basket of indicators	CTSIP	http://www.centex-indicators.org	1999

	Name of Indicator (alphabetical order)	Indicator Approach	Organizations/ Authors/Contacts	Relevant Web Sites	Year Initiated
13.	Climate Analysis Indicators Tool (CAIT)	basket of climate-relevant indicators and GHG inventories	World Resources Institute	http://cait.wri.org http://www.wri.org/project/cait	2004
14.	Climate Change Performance Index (CCPI)	composite index	Germanwatch; Jan Burck	http://www.germanwatch.org/ccpi	2006
15.	Climate Vulnerability Index (CVI)	composite index	Centre for Ecology and Hydrology, UK; Caroline Sullivan, Southern Cross University, Australia	http://www.ceh.ac.uk	2005
16.	Climate Vulnerability Initiative & Climate Vulnerability Monitor	composite & basket approach (184 countries)	DARA & Climate Vulnerable Forum	http://daraint.org/climate-vulnerability-monitor/climate-vulnerability-monitor-2010	2009
17.	Coastal Community Social Vulnerability Index (CCSVI)	composite index	S. Bjarnadottir, Y. Li and M. Stewart	http://dx.doi.org/10.1007/s11069-011-9817-5	2011
18.	Coastal Resilience Index (CRI)	composite index (community resilience self-assessment tool)	National Oceanic and Atmospheric Administration (NOAA), Mississippi-Alabama Sea Grant Consortium, & Gulf of Mexico Alliance Coastal Community Resilience Team	http://www.masgc.org/pdf/masgp/08-014.pdf http://csc.noaa.gov/criticalfacilities http://stormsmart.org http://www.gulfofmexicoalliance.org/issues/resilience.html	2008
19.	Commitment to Development Index (CDI)	composite index	Center for Global Development	http://www.cgdev.org/section/initiatives/_active/cdi	2003
20.	Consumer Price Index (CPI)	composite index	U.S. Department of Labor's Bureau of Labor Statistics (BLS)	http://www.bls.gov/cpi	1919
21.	Corruption Surveys and Indices: The annual Corruption Perceptions Index (CPI) complemented by the Bribe Payers' Index (BPI) and the Global Corruption Barometer (GCB)	The 2010 CPI ranks 178 countries (measures perceived levels of corruption as determined by expert assessments & opinion surveys)	Transparency International (TI)	http://www.transparency.org http://transparency.org/policy_research http://www.transparency.org/policy_research/surveys_indices/cpi/2010	1995
22.	Dashboard of Sustainability	dashboard	Consultative Group on Sustainable Development Indicators (CGSDI), International Institute for Sustainable Development (IISD)	http://esl.jrc.it/envind/dashbrds.htm http://www.iisd.org/cgsdi http://www.iisd.org/cgsdi/dashboard.asp	2002
23.	DataHaven	basket of 400 community indicators in 8 categories	DataHaven, New Haven, CT; a partner of the National Neighborhood Indicators Partnership; Mark Abraham & Mark Speirs	http://www.ctdatahaven.org	2003
24.	Disaster Deficit Index (DDI)	composite index	Inter-American Development Bank; Disaster Risk Management Indicators Program for the Americas, Omar D. Cardona	http://www.iadb.org/exr/disaster http://dx.doi.org/10.1111/j.1467-7717.2010.01183.x http://www.idrim.net/index.php/idrim/article/view/14	2005

	Name of Indicator (alphabetical order)	Indicator Approach	Organizations/ Authors/Contacts	Relevant Web Sites	Year Initiated
25.	Disaster Pre-paredness Index (DPi) & Resiliency Index (Ri)	composite index	David M. Simpson & Matin Katirai	http://hazardcenter.louisville.edu/pdfs/wp0603.pdf	2006
26.	Disaster Risk Index	composite index	UNEP Division of Early Warning and Assess-ment Global Resource Information Database project under a con-tract to the UNDP.	http://www.grid.unep.ch/activities/earlywarning/DRI http://www.nat-hazards-earth-syst-sci.net/9/1149/2009 http://dx.doi.org/10.1007/s11069-008-9272-0	2004
27.	Displacement Risk Index	composite index	Ann-Margaret Esnard, Visual Planning Tech-nology Lab, Florida Atlantic University's School of Urban and Regional Planning; Alka Sapat and Diana Mitsova, FAU	http://dx.doi.org/10.1007/s11069-011-9799-3 http://www.vptlab.fau.edu	2011
28.	Dow Jones Sustainability Indexes	composite indexes	Dow Jones Sustainabil-ity Indexes in collabora-tion with SAM Group Holding AG	http://www.sustainability-index.com http://www.sam-group.com	1999
29.	Drought Monitor & Drought Impact Reporter (United States)	interactive Web-based archives of drought conditions and impacts information	National Drought Miti-gation Center (NDMC), University of Nebraska–Lincoln; Donald A. Wil-hite, Mark D. Svoboda, & Michael J. Hayes	http://drought.unl.edu http://drought.unl.edu/dm http://droughtreporter.unl.edu http://dx.doi.org/10.1007/s11269-006-9076-5	1999 & 2005
30.	Ecological Foot-print (EF)	systems or accounting	Global Footprint Net-work; Mathis Wacker-nagel and William Rees	http://www.footprintnetwork.org http://www.oneplanetliving.org http://www.oneplanetvision.net	1990
31.	Ecosystem Health Monitoring Pro-gram (EHMP)	basket of indi-cators; annual report cards (2000-2010)	Healthy Waterways; Jane Hunter	http://www.healthywaterways.org http://www.healthywaterways.org/ehmphome.aspx	2000
32.	Environmental Efficiency of Well-Being (EWEB)	composite index	Tomas Dietz, Michigan State University; Kyle Knight & Eugene Rosa, Department of Sociol-ogy, Washington State University; Richard York, University of Oregon	http://www.sciencedirect.com/science/article/pii/S0049089X10002735	2009
33.	Environmental Performance Index (EPI)	composite index	Yale Center for Envi-ronmental Law and Policy and Columbia University's Center for International Earth Science Information Network (CIESIN)	http://epi.yale.edu http://sedac.ciesin.columbia.edu/es/epi	2006
34.	Environmental Vulnerability Index	composite index (50 indicators)	South Pacific Applied Geoscience Commis-sion (SOPAC) and UNEP	http://www.vulnerabilityindex.net	1999
35.	EPA Climate Change Indicators	basket of 24 indicators	U.S. Environmental Protection Agency	http://www.epa.gov/climatechange/indicators.html	2010
36.	European Envi-ronment Agency Core Set of Indi-cators (CSI)	basket of 37 indicators	European Environment Agency (EEA)	http://www.eea.europa.eu/data-and-maps/indicators http://www.eea.europa.eu/themes/climate	2003

	Name of Indicator (alphabetical order)	Indicator Approach	Organizations/ Authors/Contacts	Relevant Web Sites	Year Initiated
37.	European Innovation Scoreboard (EIS) and Summary Innovation Index (SII)	EIS is a dashboard of 20 indicators; SSI is the composite index	European Commission - Enterprise and Industry	http://www.proinno-europe.eu/metrics http://www.proinno-europe.eu/page/summary-innovation-index-0	2000
38.	European Sustainable Development Indicators	headline indicators	European Commission - Eurostat	http://epp.eurostat.ec.europa.eu/portal/page/portal/sdi/indicators http://epp.eurostat.ec.europa.eu/portal/page/portal/sdi/publications	2001
39.	Flood Vulnerability Index (FVI)	composite index	Stefania Balica & Nigel Wright	http://www.unesco-ihe-fvi.org	
40.	Florida Scorecard	dashboard & scorecard; 150+ metrics	Florida Chamber of Commerce Foundation	http://www.thefloridascorecard.com http://www.flfoundation.com	2009
41.	Gender Equity Index (GEI)	composite index	Social Watch	http://www.socialwatch.org/node/11561 http://www.socialwatch.org/taxonomy/term/527	2007
42.	Gender-related Development Index (GDI) and Gender Empowerment Measure (GEM)	composite indices	United Nations Development Programme (UNDP)	http://hdr.undp.org/en/statistics/indices/gdi_gem	1995
43.	Genuine Progress Indicator for Maryland (MD-GPI)	accounting (26 indicators)	State Government of Maryland	http://www.green.maryland.gov/mdgpi	2009
44.	Genuine Progress Indicator for Utah	accounting (22 indicators)	Utah Population and Environment Coalition (UPEC)	http://www.utahpop.org/gpi.html	2006
45.	Genuine Savings Index (Adjusted Net Savings)	accounting	World Bank	http://www.worldbank.org	1997
46.	Global Aging Preparedness Index (GAP Index)	composite index	Center for Strategic and International Studies (CSIS)	http://gapindex.csis.org http://csis.org	2010
47.	Global Aquaculture Performance Index (GAPI)	sectoral composite index	Seafood Ecology Research Group at the University of Victoria, British Columbia	http://web.uvic.ca/~gapi http://web.uvic.ca/~serg/index.html http://www.seaaroundus.org/sponsor/gapi.aspx	2010
48.	Global Climate Risk Index (GCRI)	composite index	Germanwatch	http://www.germanwatch.org/klima/cri.htm http://www.germanwatch.org/klima/cri2011.pdf	2006
49.	Global Integrity Index	composite index (300+ integrity indicators)	Global Integrity (an independent nonprofit organization tracking information on governance and corruption)	http://report.globalintegrity.org/globalIndex.cfm http://report.globalintegrity.org/methodology/whitepaper.pdf http://www.globalintegrity.org	2009
50.	Globalization Index	composite index	KOF Swiss Economic Institute	http://globalization.kof.ethz.ch	2002
51.	Happy Planet Index (HPI)	composite index	New Economics Foundation; Nic Marks & Charles Seaford	http://www.happyplanetindex.org http://www.neweconomics.org http://www.neweconomics.org/projects/happy-planet-index	2006
52.	Health Indicators Warehouse (HIW)	basket (1,119 indicators)	U.S. Department of Health and Human Services; maintained by the CDC's National Center for Health Statistics; Amy Bernstein	http://www.healthindicators.gov	2011

	Name of Indicator (alphabetical order)	Indicator Approach	Organizations/ Authors/Contacts	Relevant Web Sites	Year Initiated
53.	Holistic Eco-system Health Indicator (HEHI)	hierarchical composite indicator	Center for Sustainable Development Studies, Costa Rica; T. Muñoz-Erickson (ASU), B. Aguilar-González (Prescott College), and T. Sisk (Northern Arizona Univ.)	http://cfpub.epa.gov/ncer_abstracts/INDEX.cfm/fuseaction/display. abstractDetail/abstract/7332	1999
54.	Human Development Index (HDI)	composite index	United Nations Development Programme (UNDP)	http://hdr.undp.org/en/statistics/hdi	1990
55.	Human Security Index (HSI)	composite index (Economic, Environmental, & Social Fabric subindices; 35 indicators)	David Hastings (NOAA)	http://www.humansecurityindex.org	2008
56.	Index of Economic Well-Being (IEWB)	composite index	Centre for the Study of Living Standards; Lars Osberg and Andrew Sharpe	http://www.csls.ca/iwb.asp http://www.csls.ca/iwb/oecd.asp	1998
57.	Index of Human Insecurity (IHI)	composite index (4 categories: environment, economy, society, and institutions)	Global Environmental Change and Human Security Project, a core project of the International Human Dimensions Programme (IHDP)	http://www.gechs.org/aviso/06/ http://www.gechs.org/	2000
58.	Index of Human Progress (IHP)	composite index (10 development indicators)	Fraser Institute (an independent non-partisan research and educational organization based in Canada)	http://oldfraser.lexi.net/publications/pps/52/MeasuringDevelopmentIHP. pdf http://www.fraserinstitute.org	2001
59.	Index of Knowledge Societies (IKS)	composite index: assets, advancement, foresightedness	United Nations Online Network in Public Administration and Finance (UNPAN)	http://ictlogy.net/wiki/index.php?title=Index_of_Knowledge_Societies http://www.unpan.org http://unpan1.un.org/intradoc/groups/public/documents/un/un-pan020643.pdf	2005
60.	Index of Sustainable Economic Welfare (ISEW)	accounting	Herman Daly, John Cobb, & Clifford Cobb	http://www.foe.co.uk/community/tools/isew/make-own.html	1989
61.	Innovation Capacity Index (ICI)	composite index	World Economic Forum, The Global Competitiveness Report	http://www.innovationfordevelopmentreport.org/ici.html http://www.isc.hbs.edu/Innov_9211.pdf http://composite-indicators.jrc.ec.europa.eu/CI_Inf0002.htm	2001
62.	International Living Quality of Life Index	composite index; ranks 192 countries; 9 categories (cost of living, culture, economy, environment, freedom, health, infrastructure, safety/risk, & climate)	International Living	http://internationalliving.com http://internationalliving.com/2010/12/quality-of-life-2011 http://internationalliving.com/2010/12/quality-of-life-index-2011-where-the-numbers-come-from	1979
63.	Jacksonville's Quality of Life Initiative	basket of indicators (100+ indicators in 9 areas)	Jacksonville Community Council Inc. (JCCI)	http://www.jcci.org/	1985

	Name of Indicator (alphabetical order)	Indicator Approach	Organizations/ Authors/Contacts	Relevant Web Sites	Year Initiated
64.	Local Disaster Index (LDI)	composite index	Inter-American Development Bank; Disaster Risk Management Indicators Program for the Americas, Omar D. Cardona	http://www.idrim.net/index.php/idrim/article/view/14	2005
65.	Metropolitan Philadelphia Indicators Project (MPIP)	Basket (300+ indicators of quality of life)	MPIP is supported by the William Penn Foundation and Temple University	http://mpip.temple.edu/	2003
66.	Millennium Challenge Corporation Indicators	basket (17 indicators in FY2011)	U.S. Government Millennium Challenge Corporation (MCC)	http://www.mcc.gov http://www.mcc.gov/pages/countries/mca http://www.mcc.gov/pages/selection/scorecards	2003
67.	Mothers' Index	composite index	Save the Children	http://www.savethechildren.net/alliance/media/newsdesk/2010-05-04.html http://www.savethechildren.net/alliance/what_we_do/every_one/reports/SOWM2010_Report.pdf http://www.savethechildren.net/alliance/what_we_do/every_one/reports/SOWM2010_EXEC_SUMMARY_2010_EO_EmbargoStamped.pdf http://www.savethechildren.org/atf/cf/%7B9def2ebe-10ae-432c-9bd0-df91d2eba74a%7D/sowm2000.pdf	2000
68.	Multidimensional Poverty Index (MPI)	composite index (10 indicators; 104 countries included in 2010 MPI)	Oxford Poverty and Human Development Initiative	http://www.ophi.org.uk/policy/multidimensional-poverty-index http://www.ophi.org.uk/policy/multidimensional-poverty-index/mpi-country-briefings/	2010
69.	National Environmental Public Health Tracking Network (EPHT)	basket of indicators	Centers for Disease Control and Prevention	http://www.cdc.gov/ephtracking http://ephtracking.cdc.gov http://www.cdc.gov/nceh/tracking	2002
70.	National Index of Violence and Harm (NIVAH)	composite index (19 variables over period 1995-2003)	James Brumbaugh-Smith, Manchester College, Indiana	http://dx.doi.org/10.1007/s11205-007-9094-6 http://www.manchester.edu/links/violenceindex/	2005
71.	National Neighborhood Indicators Partnership (NNIP)	basket of indicators (34 U.S. cities)	Collaborative effort by the Urban Institute and local partners to develop and use neighborhood-level information systems in local policymaking and community building.	http://www2.urban.org/nnip http://dx.doi.org/10.1007/978-94-007-0535-/_4	1995
72.	National Well-Being Index (NWI)	composite index	Amanda Vemuri & Robert Costanza	http://www.sciencedirect.com/science/article/pii/S092180090500279X	2006
73.	Natural Disaster Hotspots	Combines hazard exposure to earthquakes, volcanoes, landslides, floods, drought, & cyclones with historical vulnerability for gridded population & GDP per unit area.	Initiated by the World Bank & Columbia University under the umbrella of the ProVention Consortium. Full list of partners & sponsors available at: www.ldeo.columbia.edu/chrr/research/hotspots/partners.html	http://www.ldeo.columbia.edu/chrr/research/hotspots http://sedac.ciesin.columbia.edu/hazards/hotspots/synthesisreport.pdf	2005

	Name of Indicator (alphabetical order)	Indicator Approach	Organizations/ Authors/Contacts	Relevant Web Sites	Year Initiated
74.	Networked Readiness Index (NRI)	composite index	World Economic Forum, Global Information Technology Report	http://www.weforum.org/issues/global-information-technology http://reports.weforum.org/global-information-technology-report	2001
75.	New Globalization Index (NGI)	composite index (21 variables)	Petra Vujakovic	http://dx.doi.org/10.1007/s11293-010-9217-3	2010
76.	Ocean Health Index (OHI) scheduled to launch February 2012	composite index (40 categories)	Conservation International, National Geographic Society, and the New England Aquarium	http://www.conservation.org/sites/marine/initiatives/ocean_health_index	2011
77.	OECD Better Life Index	composite index (11 topics; 34 countries)	Organisation for Economic Co-operation and Development, Better Life Initiative	http://www.oecdbetterlifeindex.org	2011
78.	OECD Environmental Indicators & Outlooks	basket of indicators	Organisation for Economic Co-operation and Development	http://www.oecd.org/topic/0,3699,en_2649_34283_1_1_1_1_37465,00.html	1989
79.	OECD Social Indicators	basket of indicators	Organisation for Economic Co-operation and Development	http://www.oecd.org/els/social/indicators/SAG	1982
80.	Open Budget Index (OBI)	composite index (94 countries in 2010 OBI)	International Budget Partnership	http://www.internationalbudget.org/what-we-do/open-budget-survey	2006
81.	Pay Now, Pay Later (PNPL)	basket (state-by-state assessment of the costs of climate change)	Secure American Future, a program of the American Security Project	http://www.secureamericanfuture.org/pay-now-pay-later http://www.secureamericanfuture.org	2011
82.	Predictive Indicators of Vulnerability and Adaptive Capacity	composite index	Adger et al. (2004)	http://www.tyndall.ac.uk/sites/default/files/it1_11.pdf	2004
83.	Prevalent Vulnerability Index (PVI)	composite index (24 indicators in 3 categories: exposure & susceptibility, socio-economic fragility, & social resilience	Inter-American Development Bank (IADB); Disaster Risk Management Indicators Program for the Americas, Omar D. Cardona	http://www.idrim.net/index.php/idrim/article/view/14	2005
84.	Regional Vancouver Urban Observatory (RVu) and Metro Vancouver's Vital Signs	basket of indicators	Meg Holden, Urban Studies Program, Simon Fraser University, Vancouver, BC; Vancouver Foundation	http://www.rvu.ca http://www.vancouverfoundationvitalsigns.ca http://dx.doi.org/10.1007/s11205-008-9304-x http://www.sciencedirect.com/science/article/pii/S0264275106000230	2004
85.	Resources-Infrastructure-Environment (RIE) Index	composite index	Riccardo Natoli & Segu Zuhair	http://eprints.vu.edu.au/1418 http://dx.doi.org/10.1007/s11205-010-9695-3	2008
86.	Risk Management Index (RMI)	composite index (24 indicators); measures risk management performance & effectiveness	Inter-American Development Bank; Disaster Risk Management Indicators Program for the Americas, Omar D. Cardona; Carreño, Cardona & Barbat (2007)	http://www.idrim.net/index.php/idrim/article/view/14 http://dx.doi.org/10.1007/s11069-006-9008-y	2005

	Name of Indicator (alphabetical order)	Indicator Approach	Organizations/ Authors/Contacts	Relevant Web Sites	Year Initiated
87.	Risk Reduction Index (RRI)	composite index (38 indicators)	DARA	http://daraint.org/human-impact-of-climate-change/disaster-risk-reduction-initiative	2009
88.	Social Vulnerability Index (SoVI)	composite index (32 socio-economic variables)	Hazards and Vulnerability Research Institute; Susan Cutter, Bryan Boruff, and W. Lynn Shirley	http://www.sovius.org http://webra.cas.sc.edu/hvri/products/sovi.aspx	2003
89.	Sustainable Seattle Indicators	Happiness Index (composite; 9 domains)	Sustainable Seattle (a regional sustainability indicator organization)	http://www.sustainableseattle.org http://www.sustainableseattle.org/programs/regionalindicators	1991
90.	Sustainable Society Index (SSI)	composite index (24 indicators covering human, environmental, & economic wellbeing)	Sustainable Society Foundation	http://www.ssfindex.com/data/ http://www.ssfindex.com/ssi/using-ssi/	2006
91.	Trade and Development Index (TDI)	composite index (29 indicators; 110 countries)	United Nations Conference on Trade and Development (UNCTAD)	http://www.unctad.org	2005
92.	UNEP Key Environmental Indicators	basket of indicators	United Nations Environment Programme (UNEP)	http://www.unep.org/yearbook/2011/pdfs/key_environmental_indicators.pdf	2011
93.	Virginia Performs	basket of indicators & scorecard	Council on Virginia's Future	http://vaperforms.virginia.gov http://www.statesperform.org	2007
94.	Water Poverty Index (WPI)	composite index	Centre for Ecology and Hydrology (CEH); C. Sullivan, Southern Cross Univ.; P. Lawrence, Keele Univ.; J. Meigh, CEH	http://dx.doi.org/10.1007/s11205-009-9501-2 http://gisweb.ciat.cgiar.org/wcp/download/Water_Poverty_Index_Sullivan.pdf http://129.3.20.41/eps/dev/papers/0211/0211003.pdf http://www.ceh.ac.uk	2002
95.	Water Vulnerability Index (WVI)	composite index	Caroline Sullivan, Southern Cross University, Australia	http://dx.doi.org/10.1007/s00477-010-0426-8	2011
96.	Water Wealth Index	composite index	International Water Centre, Australian Water Research Facility, University of the City of New York (CUNY); Caroline Sullivan, Southern Cross University, Australia	http://www.watercentre.org/projects/awrf-global-indicators http://nora.nerc.ac.uk/4111/	2005
97.	Well-Being Index (Well-Being Assessment)	composite index; arithmetic mean of Human Well-Being Index (36 indicators) & Ecosystem Well-Being Index (51 indicators)	Robert Prescott-Allen	http://islandpress.org/bookstore/details3d35.html?prod_id=875	2001

	Name of Indicator (alphabetical order)	Indicator Approach	Organizations/ Authors/Contacts	Relevant Web Sites	Year Initiated
98.	Well-Being Index (Gallup-Healthways)	composite index (tracks well-being of U.S. residents; congressional district, city, state, and national levels; "designed to be the Dow Jones of health")	Gallup and Healthways	http://www.well-beingindex.com http://www.well-beingindex.com/stateCongresDistrictRank.asp http://www.gallup.com/poll/106756/galluphealthways-wellbeing-index.aspx	2008
99.	World Development Indicators (WDI)	basket of indicators (900+ indicators for 213 economies)	World Bank	http://data.worldbank.org/indicator http://data.worldbank.org/data-catalog/world-development-indicators http://data.worldbank.org/data-catalog/world-development-indicators/wdi-2011	1990
100.	Worldwide Governance Indicators (WGI)	composite & basket (212 countries; voice & accountability, political stability & absence of violence, gov't effectiveness, regulatory quality, rule of law, & control of corruption)	Daniel Kaufmann, Brookings Institution; Aart Kraay, World Bank Development Economics Research Group; Massimo Mastruzzi, World Bank Institute	http://info.worldbank.org/governance/wgi/index.asp http://www.govindicators.org http://info.worldbank.org/governance/wgi/pdf/booklet_decade_of_measuring_governance.pdf	late 1990s